BLACKSMITH'S MANUAL ILLUSTRATED

A PRACTICAL TREATISE ON
MODERN METHODS OF
PRODUCTION
FOR
BLACKSMITHS, APPRENTICE
BLACKSMITHS,
ENGINEERS AND OTHERS

BY

J. W. LILLIGO

Copyright © 2013 Read Books Ltd.
This book is copyright and may not be
reproduced or copied in any way without
the express permission of the publisher in writing

British Library Cataloguing-in-Publication Data
A catalogue record for this book is available from the
British Library

Contents

Blacksmithing 1

Introductory 7

Forges Or Hearths. 11

Tools. 23

Estimation Of Lengths Of Material. 76

Welding Methods 99

Supports 105

Hardening And Tempering 215

BLACKSMITHING

A blacksmith is a metalsmith who creates objects from wrought iron or steel. He or she will forge the metal using tools to hammer, bend, and cut. Blacksmiths produce objects such as gates, grilles, railings, light fixtures, furniture, sculpture, tools, agricultural implements, decorative and religious items, cooking utensils, and weapons. While there are many people who work with metal such as farriers, wheelwrights, and armorers, the blacksmith had a general knowledge of how to make and repair many things, from the most complex of weapons and armour to simple things like nails or lengths of chain.

The term 'blacksmith' comes from the activity of forging iron or the 'black' metal - so named due to the colour resulting from being heated red-hot (a key part of the blacksmithing process). This is the black 'fire scale', a layer of oxides that forms on the metal during heating. The term 'forging' means to shape metal by heating and hammering, and 'Smith' is generally thought to have derived either from the Proto-German 'smithaz' meaning 'skilled worker' or from the old English 'smite' (to hit). At any rate, a blacksmith is all of these things; a skilled worker who hits black metal!

Blacksmiths work by heating pieces of wrought iron or steel, until the metal becomes soft enough to be shaped with hand tools, such as a hammer, anvil and chisel. Heating is accomplished by the use of a forge fuelled by propane, natural gas, coal, charcoal, coke or oil. Some modern blacksmiths may also employ an oxyacetylene or similar blowtorch for more localized heating. Colour is incredibly important for indicating the temperature and workability of the metal: As iron is heated to increasing temperatures, it first glows red, then orange, yellow, and finally white. The ideal heat for most forging is the bright yellow-orange colour appropriately known as a 'forging heat'. Because they must be able to see the glowing colour of the metal, some blacksmiths work in dim, low-light conditions. Most however, work in well-lit conditions; the key is to have consistent lighting which is not too bright – not sunlight though, as this obscures the colours.

The techniques of smithing may be roughly divided into forging (sometimes called 'sculpting'), welding, and finishing. Forging is the process in which metal is shaped by hammering. 'Forging' generally relies on the iron being hammered into shape, whereas 'welding' involves the joining of the same, or similar kind of metal. Modern blacksmiths have a range of options to accomplish such welds, including forge welding (where the metals are heated to an intense yellow or white colour) or more modern methods such as arc welding (which uses a welding power supply to create an electric arc between an electrode and the base material to melt the metals at the welding point). Any foreign material in the weld, such as the oxides or 'scale' that typically form in the fire, can weaken it and potentially cause it to fail. Thus the mating surfaces to be joined must be kept clean. To this end a smith will make sure the fire is a reducing fire: a fire where at the heart there is a great deal of heat and very little oxygen. The smith will also carefully shape the mating faces so that as they are brought together foreign material is squeezed out as the metal is joined.

Depending on the intended use of the piece, a blacksmith may finish it in a number of ways. If the product is intended merely as a simple jig (a tool), it may only get the minimum treatment: a rap on the anvil to break off scale and a brushing with a wire brush. Alternatively, for greater precision, 'files' can be employed to bring a piece to final shape, remove burrs and sharp edges, and smooth the surface. Grinding stones, abrasive paper, and emery wheels can further shape, smooth and polish the surface. 'Heat treatments' are also often used to achieve the desired hardness for the metal. There are a range of treatments and finishes to inhibit oxidation of the metal and enhance or change the appearance of the piece. An experienced smith selects the finish based on the metal and intended use of the item. Such finishes include but are not limited to: paint, varnish, bluing, browning, oil and wax.

Prior to the industrial revolution, a 'village smithy' was a staple of every town. Factories and mass-production reduced the demand for blacksmith-made tools and hardware however. During the 1790s, Henry Maudslay (a British machine tool innovator) created the first screw-cutting lathe, a watershed event that signalled the start of blacksmiths being replaced by machinists in factories. As demand for their products declined, many more blacksmiths augmented their incomes by taking in work shoeing horses (Farriery). With the introduction of automobiles, the number of blacksmiths continued to decrease, with many former blacksmiths becoming the initial generation of automobile mechanics. The nadir of blacksmithing in the United States was reached during the 1960s, when most of the former blacksmiths had left the trade, and few if any new people were entering it. By this time, most of the working blacksmiths were those performing farrier work, so the term *blacksmith* was effectively co-opted by the farrier trade.

More recently, a renewed interest in blacksmithing has occurred as part of the trend in 'do-it-yourself' and 'self-sufficiency' that occurred during the 1970s. Currently there are many books, organizations and individuals working to help educate the public about blacksmithing, including local groups of smiths who have formed clubs, with some of those smiths demonstrating at historical sites and living history events. Some modern blacksmiths who produce decorative metalwork refer to themselves as artist-blacksmiths. In 1973, the Artist Blacksmiths' Association of North America was formed and by 2013 it had almost 4000 members. Likewise the British Artist Blacksmiths Association was created in 1978, and now has about 600 members. There is also an annual 'World Championship Blacksmiths'/Farrier Competition', held during the Calgary Stampede (Canada). Every year since 1979, the world's top blacksmiths compete, performing their craft in front of thousands of spectators to educate and entertain the public with their skills and abilities. We hope that the current reader enjoys this book, and is maybe encouraged to try, with the correct training, some blacksmithing of their own.

INTRODUCTORY

In compiling this book on Blacksmith work, I have in mind the many little difficulties which arise from time to time in this class of work.

In my own experience, and also in that of my fellow workmen, problems both of time saving and labour saving have had to be solved, and the "tricks of the trade" and "wrinkles" which have been learned thereby, are passed on in this book to anyone who can make use of them. I trust that they will be found of real service to the young and ambitious smith. Blacksmithing is a trade difficult to learn. Well termed the King of Trades, practically every kindred trade depends on it in some shape or form. Tools, without which modern methods could not be developed, have to be speedily made, repaired and tempered.

I have endeavoured in this book to demonstrate, by drawings and simple text matter, specimens of smith work commonly done, and the best, simplest and quickest way to do them. From my own experience, gained at the forges of different engineering works, I have tried to pass on the easiest and best methods of arriving at the finished job.

The different types of forged work seen to-day, and the various methods by which they may be done, appear to be endless. It is not surprising, therefore, that many smiths are often at a loss as to how to commence a job and how best to proceed with it. It is no uncommon sight to see a smith commence with what should really be an intermediate or final operation. Valuable time and material is often lost through such methods.

With a view to surmounting this difficulty, I have illustrated the finished article, the commencing, following-through and final operations, which have proved under various conditions to be most successful. To become a good smith, the ability to concentrate one's mind on the work in hand is necessary. While the iron is in the fire, the smith should be mentally visualising the various operations to be gone through immediately the iron is ready.

He is a poor workman who brings his heated iron below the hammer with no clear idea in his head as to what he intends to do first. A good motto would be, "Think first and act afterwards." The smith who is well equipped with tools will often finish his job in one heat, whereas the smith using antiquated methods will require three or four heats for the same job. Some of the tools illustrated in this book might almost be called "labour-saving gadgets," as in many cases they have no resemblance to the orthodox tool. The smith who has to rely on his striker has obviously to use different methods from the smith who has the advantage of the steam hammer.

Rapid calculations plays an important part in modern smith work, and the smith who can reckon in figures the required length of material necessary to do a certain job has the advantage of his fellow workman who merely relies on guesswork. I do not suggest that the working blacksmith should be a skilled mathematician, and I have therefore embodied in this work one or two simple formulas for calculating length, which will be found to work out very well in practice. These formulas can quickly be acquired by memory, and the smith will then be saved the worry of wondering whether he has cut enough material for a job, or whether he is going to have a big waste of bar.

In a sentence, I have endeavoured to show, by illustrations and text matter, how to obtain the length of material for a job, the tools required, and the operations necessary to complete the job in the most expeditious manner.

J. W. LILLICO.

FORGES OR HEARTHS.

CAST-IRON FORGE

There are various forges in connection with blacksmith work, and the illustrations given show one or two designs in common use.

In FIG. 1 is shown a cast-iron forge fitted with a water-cooled tuyère, which protects the nose from burning when coming in constant contact with the fire.

If at any time the blacksmith's shop requires to be rearranged, this design of forge can be easily moved, not being fixed to the floor, as is the case with the brick forge illustrated on the following plate.

BRICK FORGE

In FIG. 1 is shown a common type of forge which is built of bricks. It is fitted with a water-cooled tuyère and a water trough underneath the hearth. This forge, unlike the one illustrated in PLATE 1, is a fixture and cannot be moved about.

The average height of the hearth is about 2 ft., having a length 3 ft. 6 ins. and a width 3 ft.

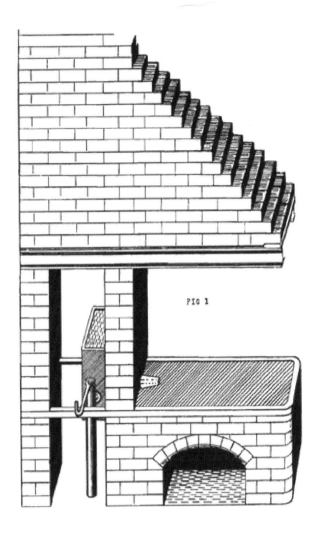

FIG 1

HEARTHS

In FIG. 1 is shown a method to adopt when heating large quantities of small tools all over. Arrange a few bricks on the hearth so as to form a small furnace, the bricks being kept together by means of wet coal surrounding them.

The fuel for such a fire is coke. Place the tools to be heated on top of the coke, and to get a good heat, place a brick in front of the fire. This can be easily moved, when taking the tools out, by sliding it along on the bricks placed for that purpose.

In FIG. 2 is shown another method for heating large quantities of tools which have only to be heated at the ends.

Take a 1½-inch square bar and double it as shown Place this bar in front of the fire and bank over with wet coal. This forces the heat through the opening, and by placing the tools between the bars a satisfactory result can be obtained.

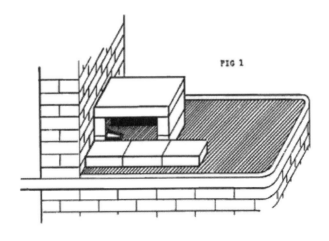

FIG 1

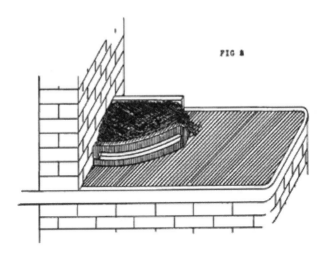

FIG 2

POT FIRE

FIG. 1 shows what is commonly called a pot fire. It can be adapted in many ways, and the building of such a fire is very simple. It is made about 2 ft. high and 3 ft. square, and should be situated so that the smith can work at it from all sides.

FIG. 2 shows a section of the pot fire. The fire hole is 18 ins. deep, 12 ins. at the base and 6 ins. at the top, the air-blast entering 15 ins. from the top. The base of the fire hole is composed of a sliding door which is easily pulled out for cleaning the fire. After the fire is cleaned, adjust the door and place on fop dead ashes, as shown in the sketch, reaching to the blast entrance. This prevents it from becoming too hot. The best fuel to use for such a fire is coke.

The Author's experience has shown this fire to have no superior in heating large forgings and in welding. Having no canopy hanging over, as in the previous illustrations, it is easy to work at.

The force of air can be increased by arranging an air-blast at the opposite side, similar to the one shown.

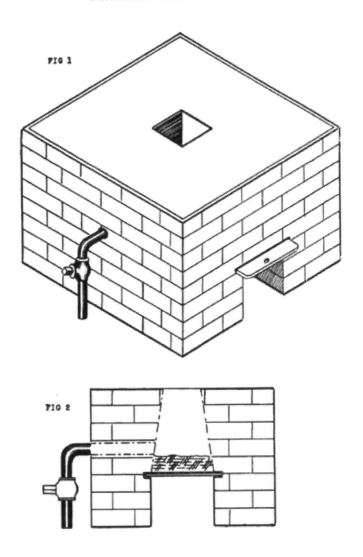

FIG 1

FIG 2

POT FIRE.

FIG. 1 shows a method of building on top of a pot fire with bricks, when heating large bars. The advantage of such a fire is that the material is heated with flame, which keeps it free from all dirt.

To build such a fire, arrange two walls of bricks in single tier three or four bricks high and place on top large flat bricks. Enclose the bar by placing loose bricks around it as shown. When the bar is heated, these can be easily removed. When heating very large bars they should rest on two bricks, one at each side of the fire hole, thus allowing the flame to circulate around the bar.

FIG. 2 shows a method of covering over the top of the fire when bricks large enough are not obtainable. Make a clam from 2-inch by $\tfrac{5}{8}$-inch bar, and slightly bend it. Bricks can be held together in this clam as shown. To lift off the clam containing the bricks, place a rod through an eye bolt which is riveted in the centre of the clam.

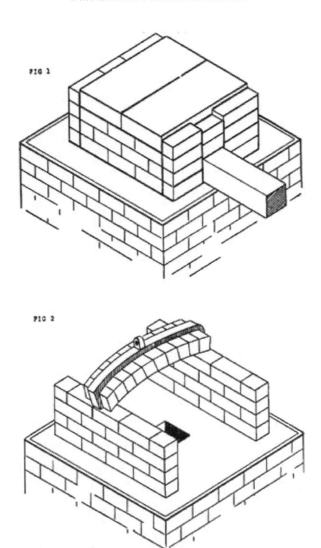

FIG 1

FIG 2

POT FIRE

FIG. 1 shows a method that is adopted for heating coal cutter picks which have to be hardened. Place bricks as shown leaving a narrow space about 30 ins. by 3 ins. through which the flame can rise. Next lay the picks on the bricks with the sharpened points over the flame. When the points become dark red, plunge into oil to harden.

This method was adopted after numerous experiments and proved the most successful.

In FIG. 2 is shown a method to heat coal cutter picks for sharpening. By arranging a few bricks to form a small furnace, the picks can be heated by placing them in and enclosing them by sliding a brick in front of the fire.

This method gives good results when large quantities of picks have to be heated. When using this method the picks are heated by the flame, and so do not burn, as sometimes occurs in an ordinary coal fire.

FIG 1

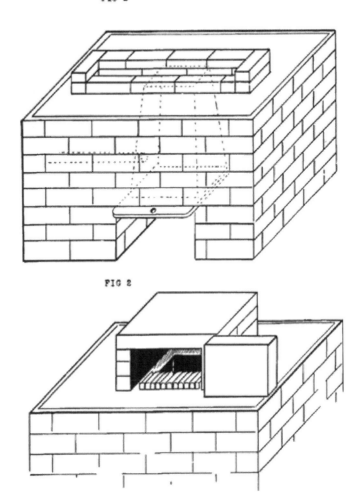

FIG 2

POT FIRE

In FIG. 1 is shown a method that can be used for carbonising when a furnace cannot be obtained. This can be arranged by placing two bricks, one on each side of the fire hole, to rest the box on containing the parts to be carbonised.

Note.—Methods of carbonising are fully explained in the Hardening chapter.

The box with its contents is cased in by building four walls around it, and covering as shown. The fire is kept burning by being occasionally filled with coke. This fuel is put in through a small opening at the front, large enough to allow a loose brick to be placed in it.

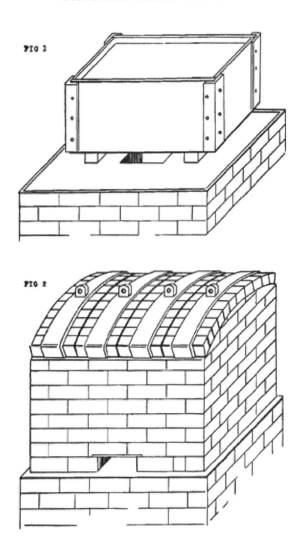

FIG 1

FIG 2

TOOLS.

ANVILS

Blacksmith's anvils. The blacksmith generally judges an anvil by its ring, a good anvil giving out a clear, sharp sound when struck with the hammer. If the anvil is defective the sound will be dull. A good anvil giving out a full volume of sound is easier to work upon than one having a dull ring. The average weight of an anvil for the smithy is 300 lb., and it is usually composed of a wrought-iron body to which is welded a hardened steel face.

To get the height in setting an anvil. The smith's finger-tips should just reach the top when standing beside it.

FIG. 1 shows an anvil set on a wood block let into the floor.

FIG. 2 shows an anvil set on a cast-iron mounting block, which is preferable, being easily moved when needed.

The practice which some smiths have of packing the anvil on their side to make it incline towards the striker is not good policy and proves a disadvantage to the smith. The anvil should be perfectly level to get the best results.

FIG. 3 shows a cast-iron swage block which is a very useful tool in the smithy. As can be seen in the illustration, the holes vary in size and shape, and around its edges are various grooves which can be selected to suit the requirements of the smith.

Blacksmith's Manual Illustrated

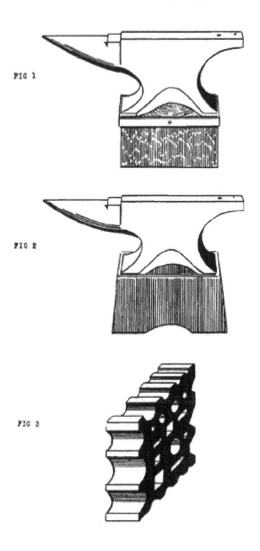

FIG 1

FIG 2

FIG 3

HAND HAMMER

In FIG. 1 is shown a hand hammer, which is made from 1½-inch octagon bar.

FIG. 2 shows the first operation by punching the hole for the shaft.

Note.—When punching a hole, it is necessary to withdraw the punch after four or five blows have been struck, so as to cool the point. A little coal dust should be put into the hole before proceeding with the punching, as this will generate gas and help to force out the punch.

Next fuller, as shown in FIG. 3, on each side of the hole, then insert the mandril and flatten on the sides, as shown in FIG. 4. FIG. 5 shows a special bolster which is used to protect the shape of the eye when driving the finishing mandril in, as shown in FIG. 6.

Another method of making a hand hammer is by working a round bar in a pair of spring tools, as shown in FIG. 7. FIG. 8 shows how the hammer is practically formed when withdrawn from the tools, leaving only the hole to be punched and the hammer cut to length.

Blacksmith's Manual Illustrated

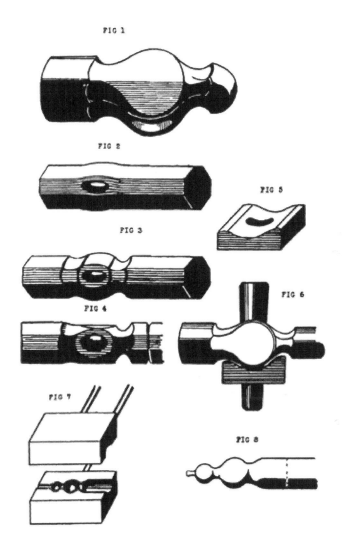

TONGS

PLATE 10 shows various kinds of tongs that are needed in the smithy, and almost every make of tongs has to be repeated to various sizes to cope with the different sections of material stocked.

Note.—Should a bar require to be forged and the tongs in stock do not fit the bar, the best remedy is to forge the end of the bar to fit the nearest size of tongs. The bar is then known to have a tong end.

This method should be encouraged as it relieves the smith in many cases from handling large and heavy tongs. For example, should the smith have to handle a piece of 4-inch square bar 12 ins. long to make a forging, instead of using 4-inch tongs, reduce the end down to 1¼-inch square or to fit tongs about that size which may be in stock. The end reduced need not be scrap when cut off, as in many cases it can be used in making smaller forgings.

FIG. 1 illustrates tongs known as pincer hollow bits. As can be seen in the illustration they are made V-shaped where they grip, thus enabling them to fit square bars as well as round bars.

FIG. 2 illustrates pincer tongs to grip between two sections as shown.

FIG. 3 shows hollow bits made to grip round or square material.

FIG. 4 shows square clip tongs also made to grip square material.

FIG. 5 shows flat tongs used with a clip, to hold various widths of material.

FIG. 6 shows duck-neb tongs used for holding hoops and bars edgeways.

FIG. 7 shows hoop tongs used for holding hoops.

FIG. 8 shows angle tongs used for gripping angle iron.

Blacksmith's Manual Illustrated

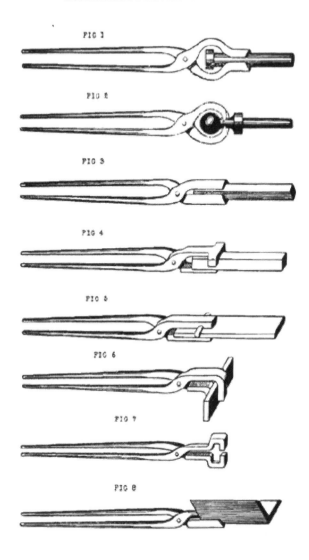

FIG 1

FIG 2

FIG 3

FIG 4

FIG 5

FIG 6

FIG 7

FIG 8

PLATE 11 deals with another collection of tongs.

FIG. 1 illustrates pick tongs which are used for holding tapered material, such as picks, etc. The rivet in the tongs in this case is an eye bolt which is used to support the tapered point of the material.

FIG. 2 shows link tongs which are used when making links.

FIG. 3 shows chisel tongs, so-called because they are used when repairing chisels.

FIG. 4 shows T-angle tongs used for holding T-bars as shown

FIG. 5 shows shingling tongs used for holding short pieces of material that have to be jumped under the steam hammer.

FIG. 6 shows pipe tongs which fit inside of the pipe.

FIG. 7 shows pliers or anvil tongs which are generally used by the assistant, to pick up odd pieces of hot material.

FIG. 8 shows rivet tongs used when riveting.

As can be seen in the illustrations the tong shanks are sufficiently open when gripping the material to enable a hand, to tighten and still avoid any possibility of the fingers being nipped in between.

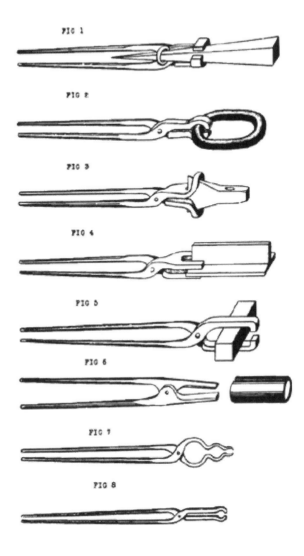

TONGS

PLATE 12 illustrates how to make tongs. When making these great care ought to be taken, as faulty tongs have been the cause of many accidents, through breaking. When making tongs avoid forging sharp angles, as these are liable to break when forging at the steam hammer, and the operator will be lucky to escape injury.

FIG. 1 illustrates the tongs to be made.

FIG. 2 shows the first operation by starting the shank of the tongs, and by following the operations shown in FIGS. 3, 4, and 5, the result is obtained.

Another method of forging tongs is to make the tong end first, as shown in FIG. 6, then follow the operations as shown in FIGS 7 and 8, and 9 which illustrates the shank ready for welding on. This method should only be adopted when there is no steam hammer available.

A method of making small tongs is to get a round bar and jump as shown in FIG. 10, then bend to shape as shown in FIG. 11. Finish off as shown in FIG. 12.

FIG. 13 shows the shape to which clip tongs should be forged.

FIG. 14 shows another method by punching a hole and cutting open, finishing by bending the ends over to fit the bar.

Blacksmith's Manual Illustrated

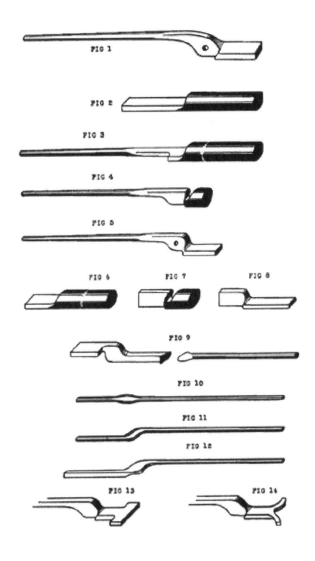

HAND TOOLS

PLATE 13 illustrates hand tools which are commonly used.

FIGS. 1 and 2 illustrate the blacksmith's hand hammer, which generally weighs between 1½ lb. and 2 lb.

FIGS. 3 and 4 show cold sets, which are used for cutting cold material and handled by a shaft.

FIG. 5 shows the same cold set, handled by a rod.

FIGS. 6 and 7 show hot sets, which are used for cutting hot material, having a finer edge than that of the cold sets.

FIGS. 8 and 9 show what is commonly called a flat face, used for levelling and finishing.

Note.—FIG. 8 has rounded edges, while FIG. 9 has square edges.

FIGS. 10 and 11 illustrate set hammers which are similar to the flat face, having round and square edges. They are commonly used, and as their name signifies, they set forgings.

FIGS. 12 and 13 show top swages used when rounding material. The eye can be punched in, either way, as shown.

FIG. 14 shows a top fuller having a straight, but rounded, edge.

FIG. 15 is a top fuller having a circular and rounded edge.

FIGS. 16 and 17 illustrate what are known as necking fullers, which can be seen in use on PLATE 47.

FIG. 18 shows a round-faced fuller.

Swages and fullers ought to be stocked in various sizes.

Blacksmith's Manual Illustrated

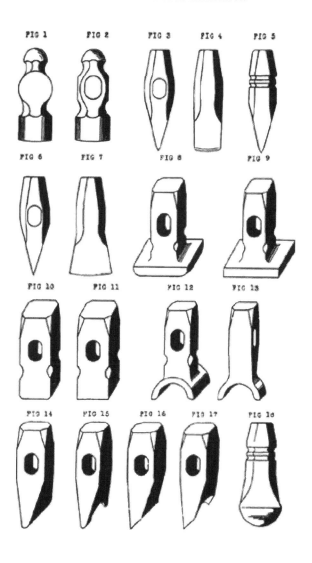

ANVIL TOOLS

It is essential that the smith be well versed in the various uses of the many tools common with his trade. The very variety of smith work itself calls for the constant necessity of some new design of tools, and the smith with an inventive turn of mind will find full scope for his ability here.

Anvil swages, as seen in FIGS. I and 2, are made in various sizes, from ¼ in. to 4 ins., with a top swage to go along with each one. They are used in rounding material to any given size that may be required. The majority of anvil tools can be made from tong ends which have been cut off the end of bars.

Note.—When making tong ends it is advisable to draw them down to suit any tool that may be required.

FIG. 3 shows a bottom fuller, which can be seen in use on PLATE 17.

FIG. 4 shows a bottom fuller with a stop forged on. This is to keep the forging from rolling off when fullering.

FIG. 5 shows a tool known as a saddle, and can be seen in use on PLATE 16.

FIG. 6 shows a bending link. See PLATE 16.

FIG. 7 shows a cutting tool which is made to come flush with the side of the anvil. This tool should be made of steel that can be hardened. See PLATE 16.

FIG. 8 shows a fork tool which is used for bending. See PLATE 16.

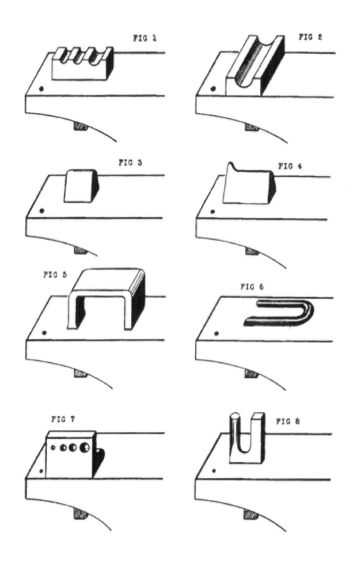

ANVIL TOOLS

FIG. 1 shows what is termed a bolt swage, so called because it is used for making bolts. It has a groove to round the bolt, and a deeper groove, which acts as a bolster when welding the head of the bolt. It also contains a hexagon groove for shaping the head as shown. A tool so made saves the trouble of changing at each operation, as is the case with single-grooved swages.

FIG. 2 shows a T-swage, so called because it is used for swaging T-pieces. See PLATE 16.

FIG. 3 shows an anvil cutter used for cutting small bars.

FIG. 4 shows a bottom radius fuller, which is explained fully on PLATE 17.

FIG. 5 shows what is termed a hexagon swage used for shaping bolt heads.

FIG. 6 shows a V-swage used for shaping square corners. See PLATE 16.

FIG. 7 shows a link tool used for welding links. See PLATE 16.

FIG. 8 shows a block tool, which is used when the anvil is too wide.

Blacksmith's Manual Illustrated

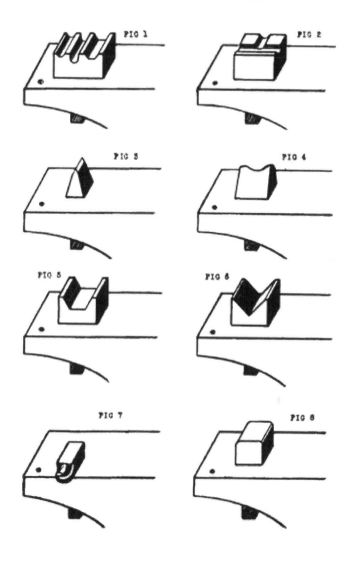

ANVIL TOOLS

PLATE 16 illustrates the tools which have previously been mentioned in use.

In FIG. 1 will be seen a saddle generally used for such forgings as double eyes.

FIG. 2 shows the method of bending a bar at right angles by using a bending link.

FIG. 3 shows the method of using the cutting tool.

FIG. 4 shows the fork tool in use.

FIG. 5 shows the bolt swage in use.

FIG. 6 illustrates a T-piece between top and bottom T-swages.

FIG. 7 shows one method of using a V-swage with a fuller to shape square corners.

FIG. 8 illustrates a method of welding a link by using a link tool.

Blacksmith's Manual Illustrated

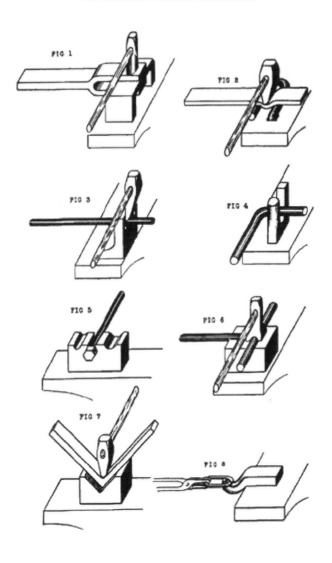

HAND TOOLS

PLATE 17 illustrates the uses of various hand tools.

FIG. 1 is showing top and bottom fullers. Fullering is generally the first operation before commencing to forge. By hammering the fullers into the depth required, as shown in FIG. 2, the smith is given a good start before drawing down, as shown in FIG. 3.

FIG. 4 shows top and bottom radius fullers, which are used for reducing a large diameter bar to a small diameter bar, as shown in FIG. 5. FIG. 6 shows the bar when drawn down.

In FIG. 7 is shown a circular chisel, which is used for cutting discs out of square plates.

FIG. 8 shows a gouge, which is used for rounding ends of bars as shown.

FIG. 9 illustrates a punch for punching holes in the material when hot. FIG. 9 also shows a sledge hammer, which is made out of 2½-inch square cast steel. The average weight of a sledge hammer is 12 lb.

FIG. 10 illustrates a cold set for cutting rivet heads off. This particular set is ground at one side only, as shown. FIG. 10 also shows a tool known as a quarter hammer, made from a 2-inch square cast-steel bar. Its average weight is 6 lb.

Blacksmith's Manual Illustrated

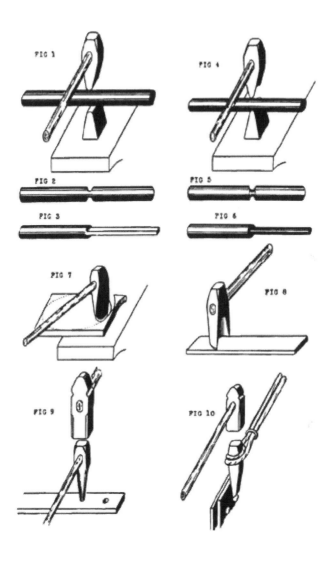

CALIPERS, ETC.

PLATE 18 illustrates a useful collection of tools very necessary for the smith in his trade.

FIG. 1 shows double calipers, which are used for checking sizes when forging.

FIG. 2 shows the handle made from a flat bar.

Note.—This has a square hole to allow each arm, as shown in FIGS. 3 and 4, to work separately when riveted. The rivet used (FIG. 5) has a square collar.

FIGS. 6 and 7 are the washers which complete its structure.

FIG. 8 represents blacksmiths' compasses or dividers.

FIGS. 9 to 12 illustrate the parts of compasses before they are riveted together.

FIG. 13 illustrates a blacksmith's T-square, which is used to square corners.

FIG. 14 shows the method of making it.

FIG. 15 represents a bevel used for setting bars to their required angles.

Blacksmith's Manual Illustrated

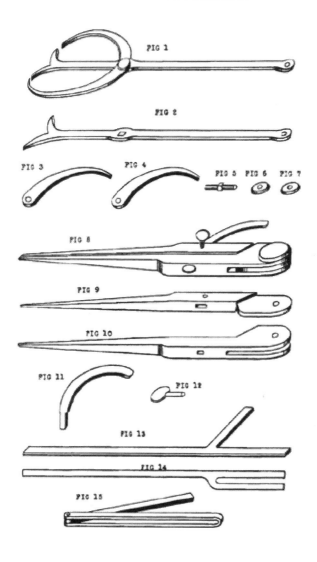

ANVIL SWAGES

PLATE 19 illustrates anvil swages. These are made in various sizes and are used for rounding material to any size that may be required.

FIG. 1 shows a short bottom swage containing three different sizes. The narrow swage is necessary for swaging between two larger sections, as in FIG. 3.

FIG. 2 represents a long single bottom swage used as in FIG. 4.

To make a swage, commence by forging 1¼-inch square from a 3-inch square bar. Next cut it off the 3-inch square bar, allowing enough material to make the body. Place in a bolster and hammer down to the required size. Next sink the groove by hammering in two or three different sizes of diameter bars, beginning with the smallest and gradually increasing the diameter to the required size.

Blacksmith's Manual Illustrated

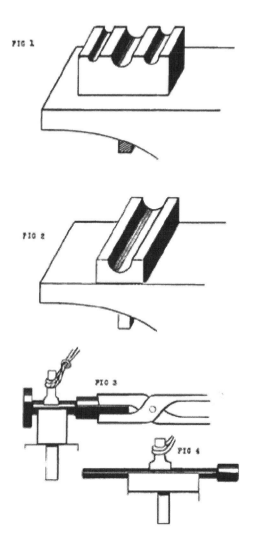

FIG 1

FIG 2

FIG 3

FIG 4

SMALL ANVIL

FIG. 1 illustrates a small anvil fixed into a large anvil, which is useful in the case of small work such as links, double eyes, etc.

FIG. 2 shows the first operation, by side-setting, 3 ins. from the end. Draw down the 3 ins. of the bar to 2 ins. diameter. Leave 4 ins. of the 2 ins. diameter, and draw down the remainder to the size of the hole in the anvil, which is about 1¼-inch square, as shown in FIG. 3. Next cut it off the bar 3 ins. long and place it in a bolster as in FIG. 4. Flatten down as in FIG. 5. Next withdraw from the bolster and hold it with the tongs, as shown, and taper the flat end down to the required size. Next draw down the opposite end and finish off.

Note.—If a smith has not a bolster high enough to hand, the difficulty can be overcome by placing several on top of each other.

Blacksmith's Manual Illustrated

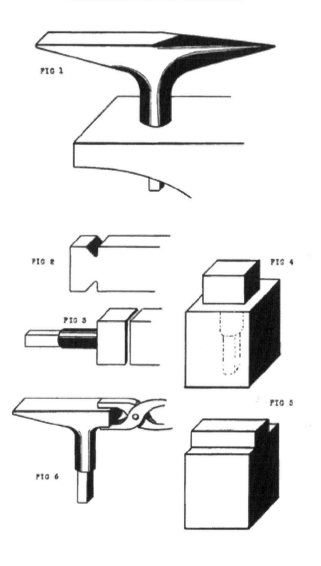

BOLSTER SWAGE

FIG. 1 shows a bolster swage which is a very useful tool, and ought to be stocked in various sizes. A large number of tools can be accumulated by using scrap pieces and shaping them while hot to any size required.

A bolster swage is similar to an anvil swage, except that it has a hole in the centre instead of a stalk.

The remainder of the illustrations on this plate show the uses to which a bolster swage can be put.

FIGS. 2 to 4 show the making of an eye bolt, before placing in a bolster swage, as in FIG. 5.

FIG. 6 illustrates a bolster swage in use during the making of a top swage.

FIG. 7 shows it in use during the making of a double eye.

FIG. 8 shows it in use during the making of a T-piece.

Blacksmith's Manual Illustrated

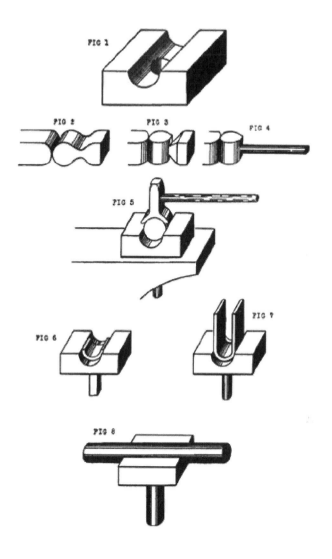

ANGLE BAR TOOLS

PLATE 22 is a collection of tools used for angle bar work. It sometimes happens that smiths have to do this class of work, although it really belongs to another branch of the trade called Angle Iron Smiths.

In FIG. 1 is shown a split block used for straighting or cutting T-bars and angle bars, as seen in FIGS. 2 and 3.

FIG. 4 shows an angle V-block suitable for welding angle bars, as shown in FIG. 5.

FIG. 6 shows a joggling block, which is used for joggling T-bars and angle bars by laying a piece of flat bar on top of the T-bar when hot under the steam hammer, and hammering down, as seen in FIG. 7.

FIG. 8 is a bending link used, as seen in FIG. 9, to bend an angle bar at right angles.

FIG. 10 shows a block to be used, as seen in FIG. 11, to bend small angle bars by gripping the bar at one end and pulling it around, at the same time hammering it on top to avoid puckering.

Blacksmith's Manual Illustrated

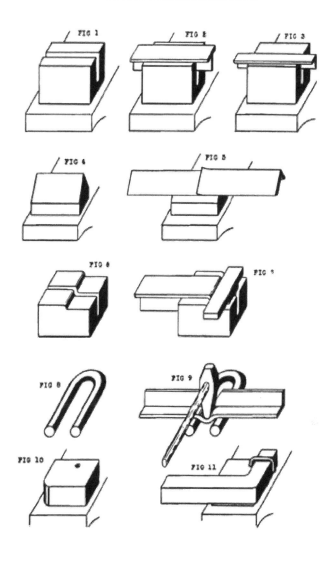

CUTTERS

PLATE 23 illustrates cold cutters for cutting cold material at the steam hammer.

FIGS. 1 and 2 show a pair of steel cutters or shears for cutting flat bars under the steam hammer. The making of a pair of cutters is very simple. Use a steel bar ¾ in. by ½ in. and cut a piece off 4 ins. long to make the bottom cutter. The top cutter has a handle forged, which is held by the smith as shown in FIG. 3.

FIG. 4 shows the position of the cutters ready to cut.

FIG. 5 shows the result.

To harden these cutters heat to a dark red, then plunge into oil. Next polish them, then lay on a hot surface till they turn dark brown.

FIG. 6 represents a cold cutter used at the steam hammer for cutting square and round bars.

FIG. 7 shows the cutter in operation.

FIGS. 8 and 9 show the method of breaking the bar when nicked around with the cutter. Place two pieces of material not exceeding 1 in. in thickness on the hammer block. Place the bar on top of them, and hold a small diameter bar on top as shown. Hit one sharp blow with steam hammer.

Blacksmith's Manual Illustrated

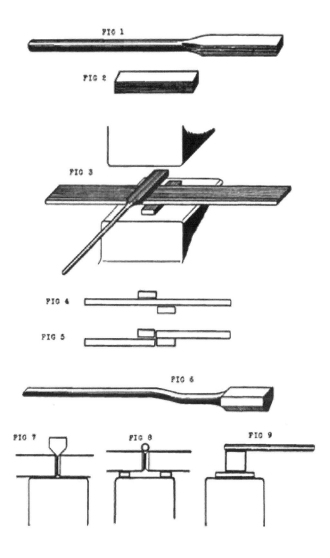

SIDE SET TOOLS

PLATE 24 illustrates various side set tools and their uses. They are mostly used for heavy forging and made in different sizes. Side sets are used similar to fullers, i.e. generally used for the first operation in forging.

In FIG. 1 is shown a pair of side sets with sharp angles, which are used to form a square shoulder when hammered in.

FIG. 2 shows a pair of side sets with the edges rounded, so as to form a radius when hammered in.

FIG. 3 represents a pair of radius side sets, which are used for side setting round or square bars.

FIG. 4 illustrates each pair of side sets hammered into the bar.

FIG. 5 shows the result.

FIG. 6 shows where only one side set has been used.

FIG. 7 shows the method of making a side set, by drawing down a handle from 3-inch square cast steel and cutting one corner off. The angle, when finished, should be perfectly square, as shown in FIG. 8.

Blacksmith's Manual Illustrated

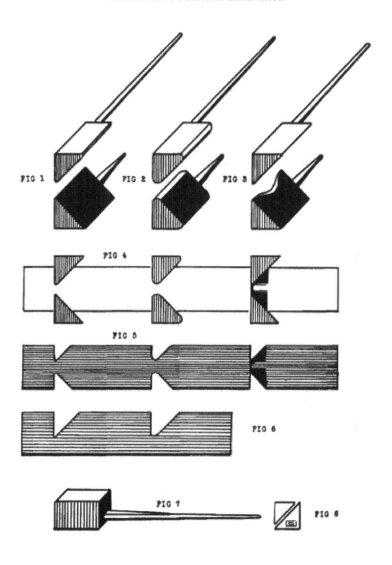

RADIUS TOOLS

PLATE 25; FIG. 1 illustrates radius cutters, which are used for cutting and rounding the end of a bar and also to form bosses.

FIG. 2 shows a pair of radius fullers, which are used to fuller and radius the bar at the same time.

FIG. 3 illustrates each pair of tools hammered into the bar, and FIG. 4 shows the result.

The method of making these tools is to hammer a square bar corner-ways into an impression which has been machined to shape, as shown in FIGS. 5 and 6.

FIGS. 7 and 8 show right-hand and left-hand cutters used for cutting hot material.

Note.—When drawing down the handles of these tools they should be made as light as possible to avoid jarring the hand.

Blacksmith's Manual Illustrated

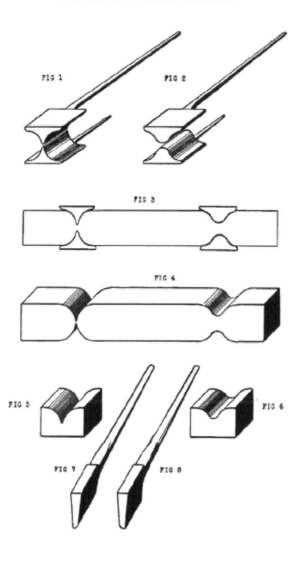

SPRING SWAGES

PLATE 26 illustrates how to make spring swages. Supposing you have to make a pair of 1-inch spring swages, as shown in FIG. I. First make two 4-inch square blocks 2 ins. thick, and in each block sink a ¾-inch diameter bar half-way down, as shown in FIG. 2. Next place a 1-inch diameter bar, as seen in FIG. 3, and sink it half-way in. When this is done, drill a ¾-inch hole in each an inch deep. Next take a ⅝-inch diameter bar, 5 ft. long, jump it at each end, then place the ends into the holes, and fix them firmly by closing in the holes with a centre punch, as in FIG. 4. Next complete the spring swages by bending the ⅝-inch bar, as shown in FIG. 1. FIG. 5 shows the swages in use.

FIGS. 6 to 10 show the making of a pair of double eye spring swages. These swages are seen in FIG. 6. The forging for which they are used is shown in FIG. 7. After making a pair of swages similar to the previous ones, stamp a flat bar, ⅝-in. thick, in one side of one swage, as shown in FIGS. 8 and 9. Repeat the same operation to the other swage. Then stamp a flat bar 1¼-in. thick between the two swages (FIG. 10).

Blacksmith's Manual Illustrated

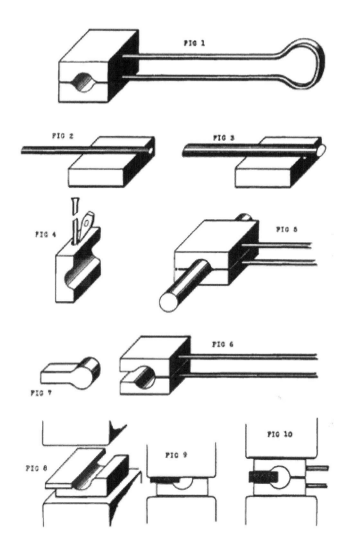

SPRING SWAGES

PLATE 27 shows a collection of useful spring tools. FIG. 1 illustrates a pair of spring swages which are used for rounding forgings to the required size. Swages for use, as in FIG. 1, should be made with the inside corners having plenty of radius, as shown. This prevents the material from sticking when revolving.

FIG. 2 shows a pair of double eye swages which really act as stamps to form double eye bosses, as shown in FIG. 3.

FIG. 4 shows a pair of middle boss swages used for stamping a boss in the middle of a bar, as shown in FIG. 5.

FIG. 6 shows a pair of eye bolt swages, used for stamping a boss with a round shank attached, as shown in FIG. 7.

FIG. 8 shows a pair of side boss swages used for stamping bosses on one side, as shown in FIG. 9.

FIG. 10 shows the method of fixing the handles in the tools, as previously described on PLATE 26.

Blacksmith's Manual Illustrated

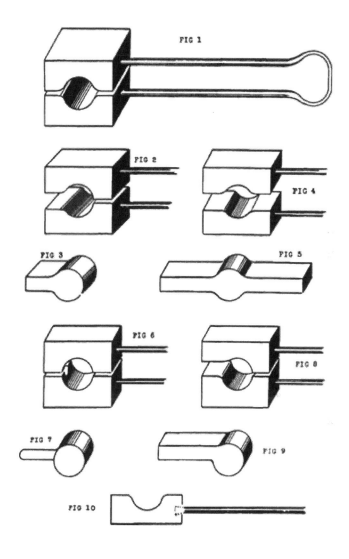

TAPERED TOOLS

PLATE 28 illustrates tapered tools and their uses. These tools are very essential in the smithy.

FIG. 1 shows a pair of tapered sets. These should be made to various angles, to enable the smith to select the required pair for any given forging that may need tapering. Should a pair of sets be required which are not in stock and it is found necessary to have them made, FIG. 2 shows the method to adopt. Mark the given taper of the required forging on a plate and make the sets to same. FIG. 3 shows the bar drawn down to the given set. Finish by cutting off at the dotted lines. These top and bottom sets are used when tapering between bosses, as shown in FIG. 4.

FIG. 5 shows a single tapered set.

In FIGS. 6 and 7 the bar is drawn down, as in FIG. 3, with the exception that in this example the bar is fullered and drawn down. The ends are then bent down, as shown in FIG. 5.

FIG. 8 illustrates the tapered set in use.

FIG. 9 shows a round tapered mandril for enlarging round holes.

FIG. 10 shows an oval mandril used for shaping the eye of a hand hammer.

FIG. 11 shows a square tapered mandril for enlarging square holes.

Blacksmith's Manual Illustrated

FIG. 12 shows a hexagon mandril used for shaping spanner jaws.

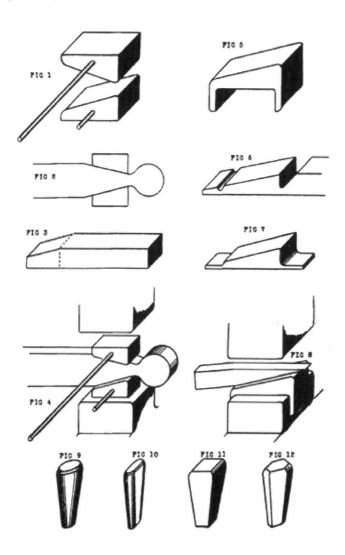

PUNCH AND DIE

PLATE 29 illustrates a very useful tool for punching holes in flat bars.

FIG. 1 shows the formation of the die. The bottom hole acts as the die and the top hole as a guide for the punch.

FIG. 2 illustrates the first operation when making a die. Draw down a 3-inch by 1½-inch bar to 3-inch by $\frac{3}{8}$-inch, then fuller between, as shown in FIG. 3, and draw down, as shown in FIG. 4. Drill a hole in each end and taper the bottom hole for clearance. Finish by bending, as in FIG. 1.

FIG. 5 illustrates the shape of the punch made of hardened steel.

FIG. 6 shows the punch and die in use.

FIG. 7 gives a sectional view of punch and die.

Blacksmith's Manual Illustrated

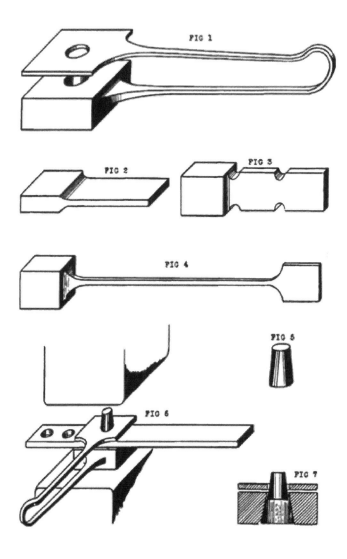

COTTER DIE

PLATE 30: FIG. 1 illustrates a very useful tool used for punching cotter holes.

FIGS. 2, 3, 4 show the method of making a cotter die for punching a ¼-inch by J-inch cotter hole. Commence by making a block of mild steel 3 ins. square, then using a tapered cotter punch, the cotter hole is punched out, FIG. 2. Next hammer a tapered mandril in to form the cotter hole, FIG. 3. FIG. 4 shows the final operation. A hole $1\frac{1}{32}$ in. in diameter is drilled in the block, to allow a 1-inch diameter bar, when hot, to enter.

FIG. 5 shows the shape of the cotter punch used when punching under the hammer.

FIG. 6 shows the punch ready to be hammered through the bar ; the result is shown in FIG. 7.

If a steam hammer is not available, an alternate method is shown in FIGS. 8, 9, 10.

FIG. 8 shows a punch being used to punch the cotter hole into the bar.

FIG. 9 shows a piece of steel being used to keep the hole in shape while swaging the bar, as in FIG. 10.

Blacksmith's Manual Illustrated

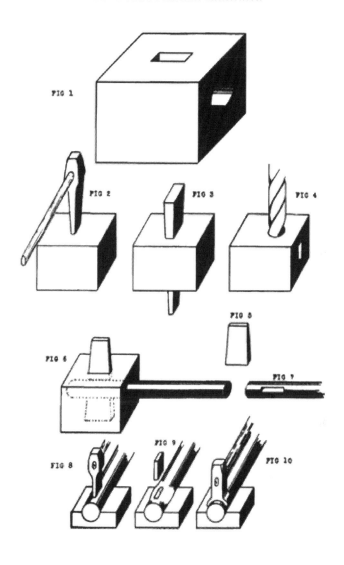

HANDLES

PLATE 31 illustrates methods of handling heaving forgings by other means than tongs.

FIG. 1 illustrates an appliance known as a portabar which is clamped to a forging. To make a portabar weld a piece of angle bar to a round bar, as shown in FIG. 2.

FIG. 3 shows a pair of handles which act as a pair of clams to clamp the portabar, when forging. In this case the striker acts in conjunction with the smith, by putting his weight on the portabar to enable the smith to turn the forging as required.

FIG. 4 shows another design of handles.

Blacksmith's Manual Illustrated

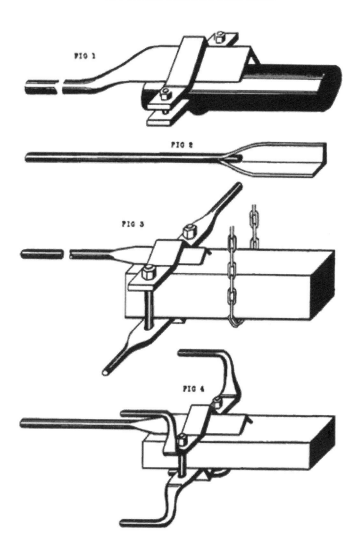

CRANE

PLATE 32 illustrates a crane supporting a forging. This apparatus is invaluable in a shop where heavy forgings are handled on the anvil and under the steam hammer. When erecting a crane in the smithy, select a position which will allow the jib to reach the fire, anvil, and steam hammer. The crane illustrated shows the jib reaching over the top of the steam hammer. It can therefore cope with any forging close to the hammer.

The following dimensions will give an idea of the material required to make a crane.

The shaft or pivot which supports the jib is 2½ ins. diameter, fixed into double eyes which are riveted or bolted to a girder, as shown.

The jib is 3 ins. by 1 in. and the tie rod 1¼ in. diameter.

On the jib two pulley wheels are held together by two plates. These support a right- and left-hand screw adjustment which in turn supports a snatch block and chain used for turning the forging.

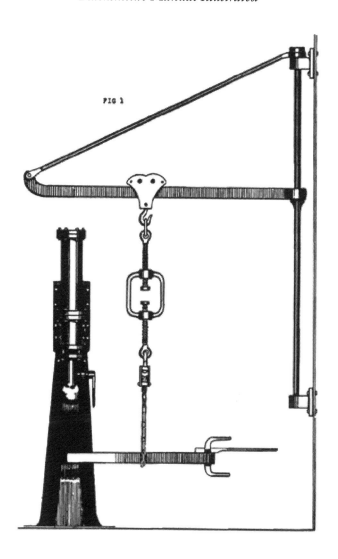

ESTIMATION OF LENGTHS OF MATERIAL

After acquainting himself with the various tools and their uses, as illustrated on the preceding plates, the young smith may find this chapter, on the estimation of lengths, of some real service to him. A great amount of unnecessary work and worry have resulted where smiths have not taken the trouble to do a simple calculation, but have endeavoured to arrive at a given result purely by guesswork.

A few examples are given here for obtaining the weight per ft. of steel bars.

To find the weight per ft. of square, round, and flat bars.

Formula : Multiply the width in eighths (inches) by the thickness in eighths (inches) and divide by 19 for square and 24 for round bars.

Examples : 6-inch square bar and 6-inch diameter bar.

```
     6 ins. sq. 6 sq.           6 ins. dia. 6 dia.
          8         8                 8         8
         ──        ──                ──        ──
         48   ×    48                48   ×    48
         48                          48
        ───                         ───
        384                         384
        192                         192
    19)2304                     24)2304
       121·2 lb. per ft.           96 lb. per ft.
```

To find the weight of flat bars adopt the same method as that for finding the weight of square bars.

Example :

```
           7½ ins. ×  3¼ ins. sq. bar
              8          8
             ──         ──
             60    ×    26
             26
            ───
            360
            120
         19)1560
            82·1 lb. per ft.
```

To find the weight per 1 in. of steel bars, multiply the width by the thickness and multiply the result by 2. Divide by 7 for square and by 9 for round bars.

Example:

$$\begin{array}{cc} 6 \text{ ins. sq.} & 6 \text{ ins. dia.} \\ \underline{6} & \underline{6} \\ 36 & 36 \\ \underline{2} & \underline{2} \\ 7\overline{)72} & 9\overline{)72} \\ 10{\cdot}2 \text{ lb. per inch.} & 8 \text{ lb. per inch.} \end{array}$$

Examples in fractions: $7\frac{1}{2}$ ins. by $3\frac{1}{4}$ ins. sq. bar.

$7\frac{1}{2}$ ins. $\times$ $3\frac{1}{4}$ ins. $\times$ $2 \div 7$

$$\frac{15}{2} \times \frac{13}{4} \times \frac{2}{7} = \frac{195}{28} = 6{\cdot}96 \text{ lb. per inch.}$$

It must be borne in mind that these methods give approximate results only, but from experience they have been found to give satisfaction.

The usefulness of arriving at these weights may not at first be apparent to the young smith, but it will be seen in examples to follow that where the weight per ft. of a bar to be forged is known and where the weight per ft. of the bar from which it is to be made down is known, a simple deduction will quickly guide the smith as to the correct length of material he requires.

FORGING

PLATE 33 gives a few examples of material to be drawn down. Find the weight of each section by following the instructions given, then from these weights the length to be drawn down can easily be found.

From the method of calculating lengths of forgings given in preceding chapters, an allowance in weight is given in practically all cases to allow for a certain amount of depreciation in the material when it is heated. The correct amount to allow will only be found with practice. To know the weight required is the chief help, but the allowance must not be ignored, otherwise the smith may find himself short of material on completion.

FIG. 1 shows 18 ins. of 1 in. diameter drawn down from 2 ins. diameter.

FIG. 2 shows the 2-inch diameter bar fullered 4½ ins. from the end required to be drawn down.

FIG. 3 shows 18 ins. of 1 in. square drawn down from 2 ins. square.

FIG. 4 shows the 2-inch square bar fullered 4½ ins. from the end required to be drawn down.

FIG. 5 shows 15 ins. of 1½ in. square drawn down from 3¾ ins. diameter.

FIG. 6 shows the 3¾-inch diameter bar fullered 3¼ ins. from the end required to be drawn down.

FIG. 7 shows 20 ins of 1½ in. diameter tapered to ¾ in. diameter drawn down from 3-inch diameter bar.

FIG. 8 shows the 3-inch diameter bar fullered 3 ins. from the end required to be drawn down.

Blacksmith's Manual Illustrated

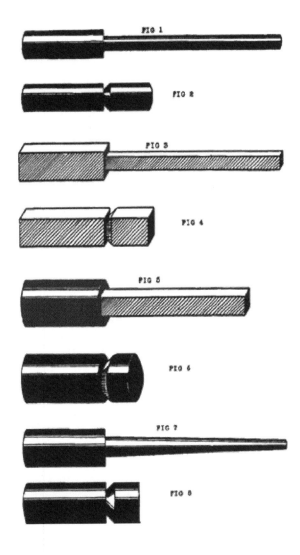

FORGINGS

PLATE 34 gives a few examples of jumping small sections into larger sections.

FIG. 1 illustrates a blank flange 6 ins. by 1 in. made from a 3-inch diameter bar. To obtain the length of 3-inch diameter bar required, calculate as follows :—

Method.—To find the weight of 6-inch diameter bar 1 in. thick. Square the diameter, multiply by 2 and divide by 9, viz. : $6 \times 6 \times 2 \div 9 = 8$ lb. per inch.

Similarly find the weight of the 3-inch diameter bar: $3 \times 3 \times 2 \div 9 = 2$ lb. per inch.
$8 \div 2 = 4$ ins.

FIG. 2 shows 4 ins. of 3-inch diameter bar which is required to make the blank flange 6 ins. diameter 1 in. thick.

FIG. 3 illustrates a forging 9, ins. diameter by 1½ in. thick made from 4-inch diameter bar.

FIG. 4 shows 7¾ ins. of 4-inch diameter bar, the quantity required to make the forging 9 ins. diameter and 1½ in. thick.

FIG. 5 illustrates a forging 7½ ins. square by 1¾ in. made from a 4½-inch square bar.

FIG. 6 shows 4½ ins. of 4½-inch square bar, the quantity required to make the 7¼ ins. by 1¾ in. forging.

FIG. 7 illustrates a wedge-shaped forging 10 ins. long by 4 ins. square at one end, made from a 4-inch square bar.

FIG. 8 shows 5 ins. of 4-inch square bar, the amount required to make the wedge-shaped forging 10 ins. long by 4 ins. square at one end.

Blacksmith's Manual Illustrated

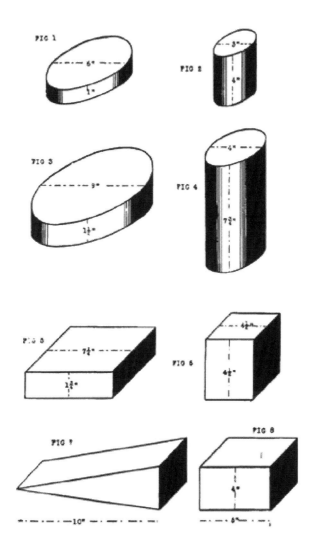

FORGING

PLATE 35 : FIG. I shows 18 ins. of 4 ins. tapered to 2 ins. by 1 in. thick from a 4-inch square bar.

FIG. 2 shows the 4-inch square bar fullered 3| ins. from the end which is required to be drawn down.

To get the best results when drawing a forging down, take a little at a time. The method to adopt is given in FIG. 3.

FIG. 4 illustrates the finishing of the taper by using tapered sets under the steam hammer.

Note.—An easy method in calculating the weight of a tapered bar is by adding the largest and smallest widths of the tapered section together and dividing by 2, thus obtaining the mean width.

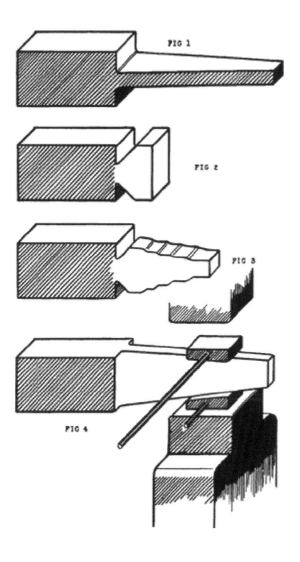

LINK. SHACKLE. CLAMS.

PLATE 36 gives a few simple methods of obtaining lengths.

FIG. 1 shows a link with an overall length of 6 ins. and an inside width of 2 ins. The length of material required to make this link is 14 ins. (FIG. 2). Method of finding length : Add twice the overall length to the inside width.

FIG. 3 shows a shackle. Length from the centre of the holes to the inside of the shackle, 6 ins. Width between, 2 ins. The length of the shackle from centre to centre of the holes before bending is 14 ins. (FIG. 4). This length is found by adding twice the given length to the inside width.

FIG. 5 shows another shackle of a different shape, its length before bending being 16 ins. from centre to centre of the holes (FIG. 6). This length is found by adding twice the distance between the centre of the holes and the centre of the opposite side, as shown, to the inside diameter of the shackle.

FIG. 7 shows a pair of 12-inch clams. An easy method of obtaining the approximate length of material required to make the bend of a half clam, is by adding the given diameter to the radius, e.g. let the diameter be 12 ins. The radius is therefore 6 ins., giving the required length 18 ins. (FIG. 8).

Blacksmith's Manual Illustrated

Note.—These lengths are approximate, and are very suitable for shop practice.

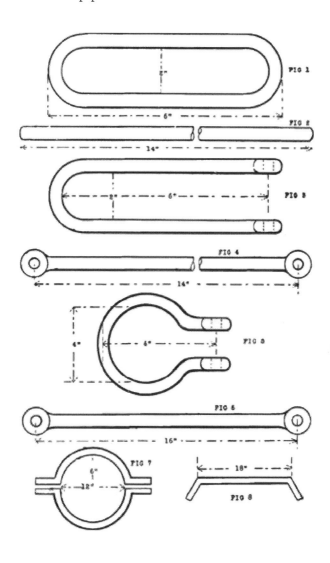

HOOP

PLATE 37 : FIG. 1 illustrates a hoop which has an inside diameter of 12 ins. and made from 3-inch by 1-inch bar. The required length of material for making such hoops often presents itself as a difficulty to the smith.

A very useful rule to follow, and one often applied in shop practice is here given. To three times the inside diameter add three times the thickness and allow ½ in. to every foot of the inside circumference. This rule, applied to the given hoop which is 12 ins., would be 40¾ ins. Applying the supposedly correct method (diameter by 3·1416 plus three times thickness and ½ in. allowance for welding) to the above example, the length required would be $41\frac{1}{8}$ ins., the difference being $\frac{3}{8}$ in. The smith would do well to follow the former method, as it is easier to draw a hoop which is too small to the correct size, than to jump one which, on completion, is too large.

FIG. 2 shows a special tool for placing hoops on for welding.

FIG. 3 shows a method of bending a bar to form a hoop.

One method for rounding a hoop is by heating it all over, placing it on a cone, and hammering it to the required shape, as shown in FIG. 4.

Blacksmith's Manual Illustrated

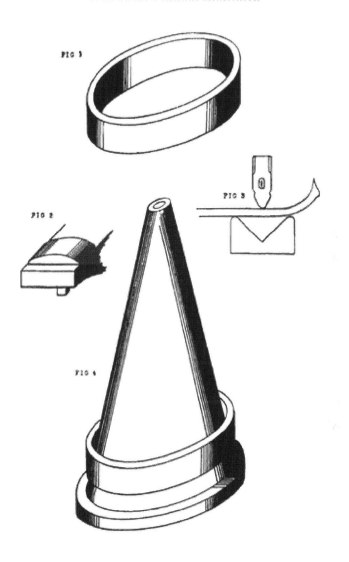

CONED HOOPS

PLATE 38: FIG. 1 shows a cone-shaped hoop, made from a 2-inch by ½-inch bar having a top inside diameter of 16 ins. and a bottom inside diameter of 18 ins.

The required length of the material (circumference) can be found by adding the two inside diameters together and dividing by 2 (mean diameter). The result (17) is then multiplied by 3·1416 which gives 53·4072 ins. To this figure should be added three times the thickness of the hoop (1½ in.) plus ½ in. for welding, making the answer 55·4072.

FIG. 2 shows the bar cut to the required length.

FIG. 3 shows the same scarfed.

The smith should remember that when making cone-shaped hoops, the material must always be bent edgeways to a given radius.

FIG. 4 shows the method of obtaining the necessary radius before bending. Lay two parallel lines the width of the bar apart with a centre line running at right angles. On the parallel lines mark off from the centre line half the diameters of the hoop, i.e. 8 ins. and 9 ins. is shown. Draw two diagonal lines through these points to touch the centre line. This gives the centre point for the compasses to draw in the inner and outer circumferences as shown. Next bend the bar to the given radius.

FIG. 5 shows the bar cambered- right for making the hoop.

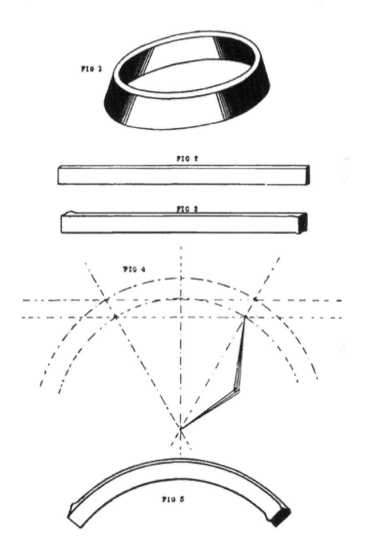

BEVEL CLAM

PLATE 39 : FIG. 1 illustrates a bevel clam with top inside measurement 6 ins., and bottom inside measurement 8 ins. The following illustrations show the method of making it. The procedure is similar to that in making a coned hoop, i.e. the bar must be bent edgeways.

FIG. 2 shows the straight bar marked 6 ins. and 8 ins. also showing the correct angle A to which the bar should be bent edgeways.

FIG. 3 shows the bar bent edgeways.

FIG. 4 shows the bar bent along the line B-C.

Blacksmith's Manual Illustrated

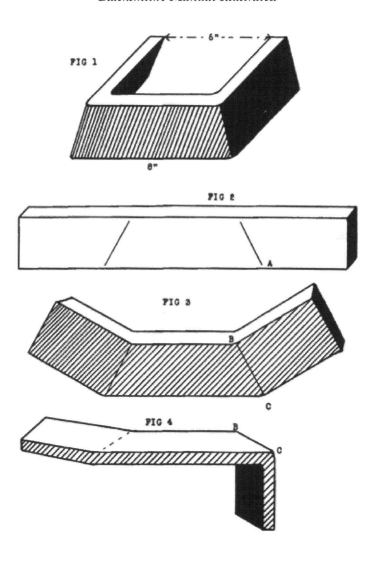

ANGLE BAR RINGS

PLATE 40 : FIG. 1 illustrates an angle bar ring with the flange bent on the inside.

FIG. 2 gives the size of the ring which is 20 ins. When the flange is bent on the inside the ring is usually measured to the extreme diameter, as shown. The method of ascertaining the length is as follows : from the diameter 20 ins., subtract twice the diagonal thickness 1 in. at A, and multiply the answer by 3·1416. 20 ins. — 2 ins. = 18 ins. 18 ins. × 3·1416 = 56·5 ins.

FIG. 3 illustrates the flange on the outside. The method to adopt in this case is to take the interior diameter, add twice the diagonal thickness and multiply their sum by 3·1416.

FIG. 4 gives the interior size, 20 ins. Method of ascertaining the length is as follows : 20 ins. + 2 ins. = 22 ins. 22 ins. × 3·1416 = 69 ins.

FIG. 5 illustrates method of bending small angle bars. Supposing the angle bar to be bent is 2 ins. by 2 ins., take a bar of 2 ins. square and bend it to the same radius that is required on the angle bar, then bend one end to fit the hole in the anvil as shown. When the angle bar is hot, grip the end as shown and pull around against the 2-inch square bar.

Blacksmith's Manual Illustrated

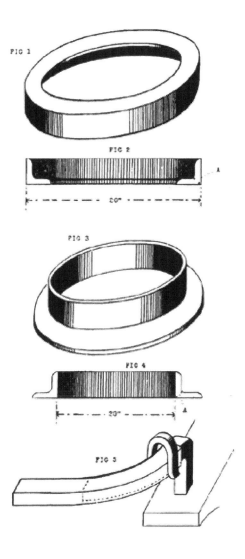

SIDE SETS

PLATE 41 illustrates side sets being used for making various forgings.

FIG. 1 illustrates the side set in use.

FIG. 2 illustrates the result before being drawn down.

FIG. 3 illustrates the result after being drawn down.

FIG. 4 illustrates two side sets in use.

FIG. 5 illustrates the results before being drawn down.

FIG. 6 illustrates the results after being drawn down.

FIG. 7 illustrates how to get a smaller section into the centre of the bar, by using two side sets one above the other.

FIG. 8 illustrates the results before being drawn down.

FIG. 9 illustrates the results after being drawn down.

FIG. 10 illustrates the bar being side set on the four sides.

FIG. 11 illustrates the result before being drawn down.

FIG. 12 illustrates the result after being drawn down.

To side set a round bar it is advisable to recess it when hot by placing a small diameter bar on top, and keep turning the large bar until it is practically fullered to the depth of the small diameter bar, as shown in FIG. 13.

FIG. 14 illustrates the placing of the side sets in the recess and continually turning the bar.

FIG. 15 illustrates the results.

FIG. 16 illustrates the method that is adopted when fullering a flat bar, by means of a round bar being hammered in.

FIG. 17 illustrates the result before being drawn down.

FIG. 18 illustrates the result after being drawn down.

FIG. 19 illustrates a round bar doubled to the required width and hammered in.

FIG. 20 illustrates the result similar to FIG. 6.

Blacksmith's Manual Illustrated

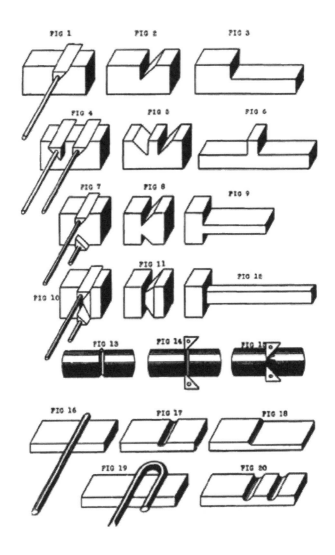

WELDING METHODS

Welding is the combining together of two pieces of iron or mild steel. This is done by heating both pieces to a plastic state, then hammering one into the other so as to form one solid bar. During the heating process it is necessary that the healed iron should be kept from coming in contact with the air-blast, the reason for this being that heated iron absorbs oxygen, thus forming a scale and preventing a good weld. Sand is a very good flux when welding mild steel. It melts and covers the heated surface, thus protecting it from oxidation and therefore is of great assistance in making a good weld.

FIG. 1 illustrates a scarf weld commonly used.

FIG. 2 illustrates a fork and wedge weld used for welding steel into iron.

FIG. 3 illustrates a butt weld which is used when one end is close up to a shoulder.

FIG. 4 illustrates a V-weld suitable for heavy rings.

FIG. 5 illustrates a stud weld, i.e. a round bar welded into a square bar forming a T-piece.

FIG. 6 illustrates another stud weld.

FIG. 7 illustrates a rivet weld suitable when making large hoops.

FIG. 8 illustrates a scarf weld suitable for small rings.

FIG. 9 is showing FIG. 1 welding under the hammer.

FIG. 10 is showing FIG. 2 welding under the hammer.

FIG. 11 is showing FIG. 4 welding under the hammer.

FIG. 12 illustrates a method of scarfing under the steam hammer.

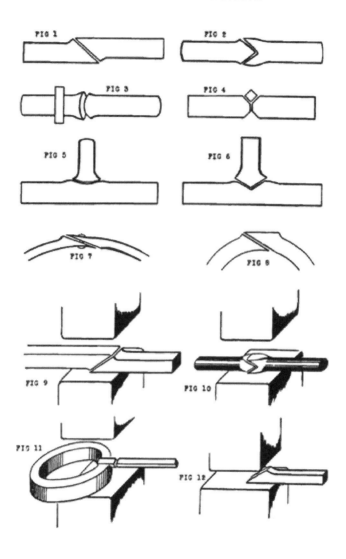

JUMPING

PLATE 43 gives a few examples of jumping. This process is used to increase the diameter, width, or thickness, at the same time reducing the length.

FIG. 1 shows a heavy flat bar turned up at each end used as an apparatus for jumping. The method of using this bar is shown. By heating the bar to be jumped, and then bending it, it can be placed between the two uprights and hammered where it is hot, thus jumping it to the required length. If the uprights are too far apart to enable the bar to be jumped between them, this can be remedied as shown.

FIG. 2 shows a method of jumping, by gripping the bar while hot under the steam hammer, and striking it with a ram, which is suspended by a chain from any convenient beam overhead.

FIG. 3 shows a pin bolt. Its head is 6 ins. diameter and 1 in. thick; its length, excluding the head, is 6 ins. having a diameter 4 ins. To make this pin bolt out of a 4-inch diameter bar, $2\frac{3}{8}$ ins. must be allowed for the head alone, making the total length of a 4-inch diameter bar required, $8\frac{3}{8}$ ins.

FIG. 4 shows the end of the 4-inch diameter bar being jumped under the steam hammer. The jumping is improved and better start given to the head by holding a flat bar on the 4-inch diameter bar, where the hammer strikes.

FIG. 5 shows another method of making a pin bolt by placing the 4-inch diameter bar in a bolster which has a recess for the head, and hammering down to form the head in the recess.

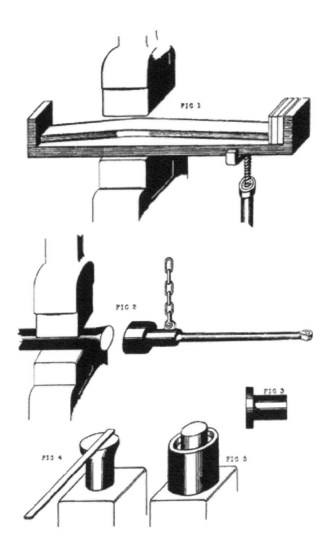

SUPPORTS

PLATE 44 illustrates various supports or adjusting stands necessary for supporting bars which are too long for the smith to handle.

FIG. 1 shows a stand which can be adjusted by placing a rod through the holes.

FIGS. 2 to 6 show the parts by which the structure of the stand shown in FIG. 1 is composed.

FIG. 7 shows another type of stand which can be easily adjusted by screwing up or down.

The stand in FIG. 8 is similar to the one shown in FIG. 7, except that when adjusted it is held by a set screw.

Blacksmith's Manual Illustrated

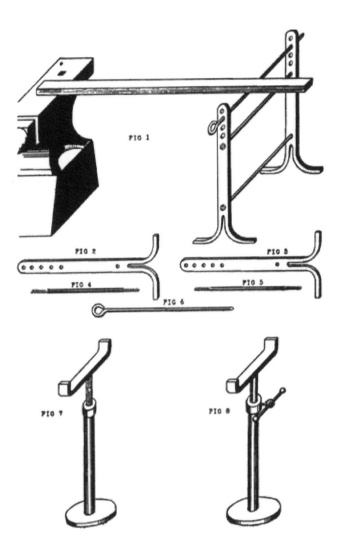

EASY METHODS

PLATE 45 illustrates a few easy methods for obtaining quick results with small material.

FIG. 1 shows a four-cornered clam made by heating the bar and laying it across a recessed tool. Then, by placing a bar across, as shown in FIG. 2, one blow under the steam hammer gives the result seen in FIG. 3.

FIG. 4 shows a clam.

FIG. 5 shows an arrangement for making clams, using two bars edgeways, which are regulated by four adjusting bolts to suit the size of clams to be made. To make the clam, lay the hot bar across the two bars and hammer the top tool down as shown.

FIG. 6 shows another method of making clams by using a recessed tool, but in this case one size of clam only can be made.

FIG. 7 shows a joggled bar, the method of joggling being shown in FIGS. 8 and 9.

FIG. 10 shows a pipe hanger, which can be made by first making a template around which the pipe hangers can be shaped. These operations are shown in FIGS. 10 to 12.

FIG. 13 shows a flat bar with two bends close together. FIG. 14 shows a bar fixed in the anvil with the fiat bar laid across, gripped with the tongs, and the ends hammered over, as shown in FIG. 15.

Note.—These methods are adopted when numerous jobs of the same type have to be made.

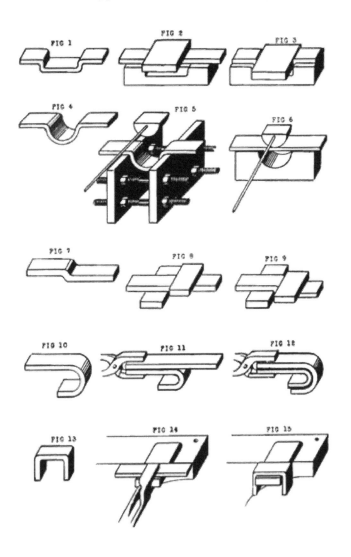

SPLIT COTTERS

PLATE 46 illustrates several ways of making split cotters.

FIG. 1 shows a split cotter.

FIG. 2 shows a square bar cut through, and flattened down as in FIG. 3; next form the head as shown in FIG. 4, then finish to size, and cut off.

Another method is by pointing the end of the bar and flattening it, as shown in FIGS. 5 and 6, then doubling it as in FIG. 7, and shaping it as in FIG. 8. Next cut it off the round bar.

A third method is by using a flat bar which has been reduced, as shown in FIG. 9. Cut it along the dotted line and place the pieces ready to weld, as shown in FIG. 10.

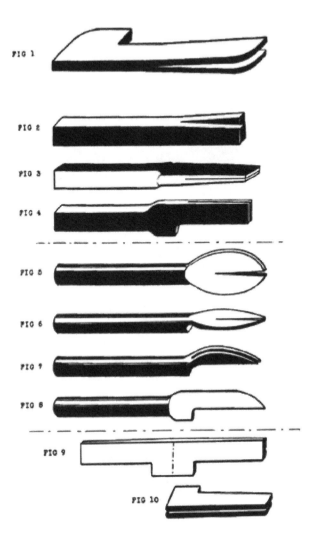

BOLTS

PLATE 47: FIG. 1 illustrates a bolt, the other FIGS. showing how it is made.

FIG. 2 shows the fullering of a hexagon bar with necking fullers. The bar is then drawn down to size, again use necking fullers, this time to round off the corners. Cut it off the bar, as in FIG. 3.

FIG. 4 shows a method of squaring the head of the bolt by placing it in a hand bolster. FIG. 5 shows the head of the bolt being corrected in an anvil swage.

FIG. 6 shows a small pin bolt, which can be made in FIG. 7, bolster if a round body is required, and in FIG. 8. bolster if a square body is required.

FIG. 9 shows a round bar in a bolster, having enough material to form the head by hammering it to fit the recess in the bolster.

Blacksmith's Manual Illustrated

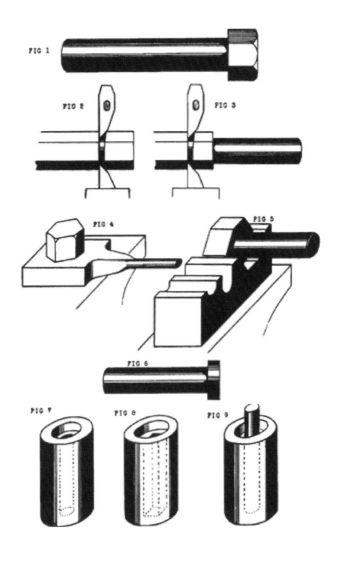

WRENCH

PLATE 48 : FIG. 1 illustrates a wrench to fit a 4-inch square nut, and made from a 2-inch by ½-inch bar.

First operation, FIG. 2: Point the ends of a 2-inch by ½-inch bar, 32 ins. long as shown.

Second operation, FIG. 3: Double over the ends as shown, making the bend 4 ins. from each end.

Third operation, FIG. 4: Bend the double ends over as shown.

Fourth operation, FIG. 5: Weld the double ends as shown, leaving the length A-B 16 ins.

Fifth operation, FIG. 6: Fuller as shown, making each division 4 ins.

Sixth operation, FIG. 7 : Bend the bar as shown.

Seventh operation : Weld the two ends together, then complete by welding a 1-inch diameter bar to it.

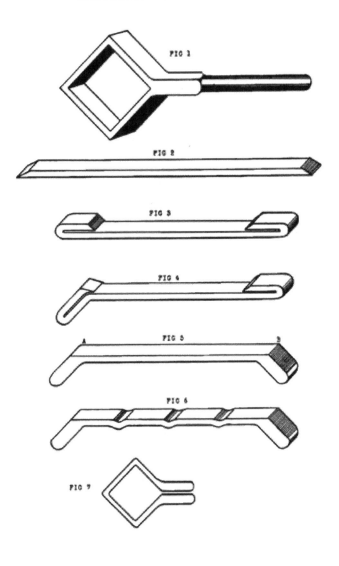

BRAKE GUIDE

PLATE 49 : FIG. 1 illustrates a brake guide, made from a 2-inch by ½-inch bar.

First operation, FIG. 2 : Bend, then weld as shown at A, leaving enough material to complete the other necessary operations.

Second operation, FIG. 3: Bend the bar at B, after finishing off A.

Third operation, FIG. 4: Bend the other end of the bar at C.

Fourth operation, FIG. 5 : Bend at D.

Fifth operation, FIG. 6: Fuller at E as shown, then bend at E, making D-C at right angles to the bottom line. Weld together, as shown in FIG. 1.

FIG. 7 illustrates a method of bending a bar by fullering, as first operation.

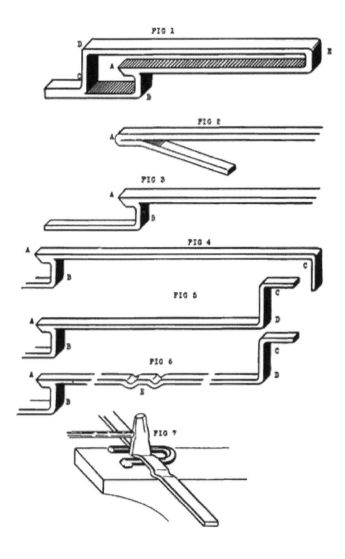

HOOKS

PLATE 50 : FIG. 1 illustrates a hook.

First operation, FIG. 2: In a flat bar make a hollow space. This is done by using a round-faced fuller, often called a bob-punch.

Second operation, FIG. 3 : Take a round bar, jump the end, and weld it into the space previously prepared. Immediately turn it over, place in a bolster (FIG. 4) and hammer to complete the welding.

FIGS. 5 to 8 illustrate another method by which this hook can be made. Punch a hole through the bar, as shown in FIG. 5. FIG. 6 shows a pin placed through the hole and riveted to the bar. FIG. 7 shows the shape of the pin which is used. Raise the bar and pin to a welding heat, then place in a bolster to complete the welding, as shown in FIG. 8.

FIG. 9 shows the bar fullered ready to draw down to the required size.

Another method of making hooks is to make them from a solid bar. Draw down, as shown in FIG. 10, leaving enough material to form the pin.

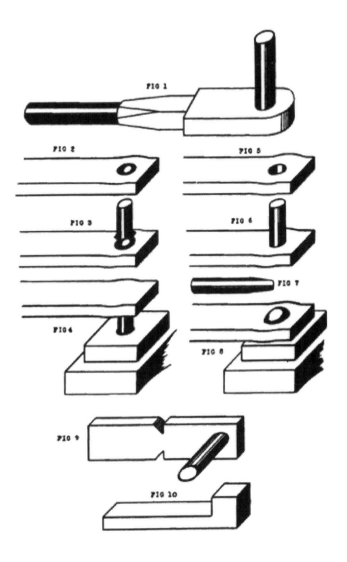

DOUBLE EYE

PLATE 51: FIG. 1 illustrates a double eye. The most satisfactory way to make this forging is as follows :—

First operation, FIG. 2: Make a recess in the centre of the bar as shown.

Second operation, FIG. 3: Place a bar to fit the recess to prevent it from altering when the bar is being fullered or side setted as shown.

Third operation, FIG. 4: Draw down the ends as shown, and cut through the dotted lines to round off the inside corners, then bend the ends as in FIG. 1. Finish off by placing on a mandril, as shown in FIG. 5.

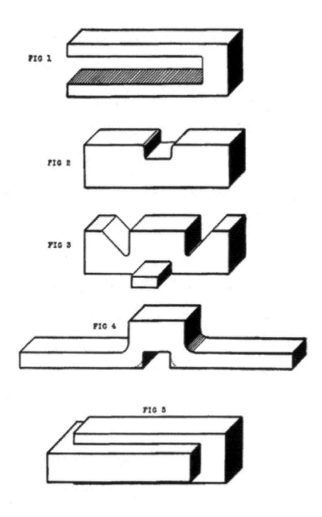

LIMBER DOUBLE EYE

PLATE 52: FIG. 1 illustrates a limber double eye and V-piece, made from a 3½-inch by 1½-inch bar.

First operation, FIG. 2: Draw down 4 ins. of the 3½-inch by 1½-inch bar to 1½ in. square as shown.

Second operation, FIG. 3: Swage the 1½-inch square to 1½ in. diameter as shown, leaving enough 1½ in. square to form the double eye.

Third operation, FIG. 4: Flatten the 1½ in. diameter as shown, and roughly shape the double eye.

Fourth operation, FIG. 5: Stamp the double eye as shown.

Fifth operation, FIG. 6: Punch a hole as shown.

Sixth operation, FIG. 7: Cut open from the end to the hole on a shallow swage; this prevents the eye from going out of shape as shown.

Seventh operation, FIG. 8: Finish off the double eye by placing a mandril in between, and hammer down under the steam hammer to the required size.

Eighth operation, FIG. 9: Start the opposite end, punch a hole in as shown, then open it out.

Ninth operation, FIGS. 10 to 11: Draw down each end to size, under the steam hammer.

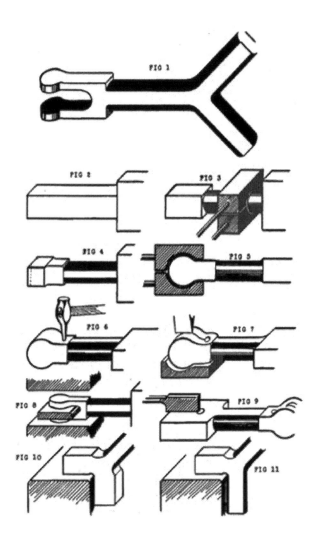

LEVER DOUBLE EYE

PLATE 53: FIG. 1 illustrates a lever double eye, made from a 5-inch by 2-inch bar. An easy method of making this is given in the following illustrations:—

First operation, FIG. 2: Side set the bar as shown, leaving, enough material in the centre to make the shank and boss, as shown in FIG. 1.

Second operation, FIG. 3: Draw down each side to the thickness of the bosses, then fuller or side set as shown.

Third operation, FIG. 4: Draw down in between the bosses, and fuller the inside comers (A) as shown. This prevents the chisel marking the drawn-out parts when shaping the bosses by cutting along the dotted lines.

Fourth operation, FIG. 5: Bend to shape, then fuller as shown, leaving enough material to form the end boss.

Fifth operation, FIG. 6: Draw down, using two narrow blocks as shown, then shape the boss as in FIG. 4.

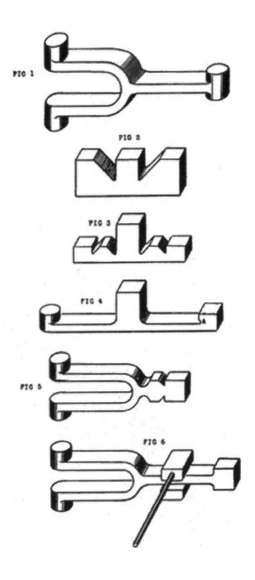

LEVER ARM

PLATE 54: FIG. 1 illustrates a lever arm, made from 3 ins. square.

First operation, FIG. 2: side set the bar as shown.

Second operation, FIG. 3: Draw down to the size of the double eye, then fuller as shown.

Third operation, FIG. 4: Draw down as shown, then cut from the bar.

Fourth operation, FIG. 5: Bend it, thus making it easier to draw out the pin. (Which is the fifth operation, FIG. 6.)

Sixth operation, FIG. 7: Straighten it, then place the pin into a bolster and hammer it level, as shown. Next shape the end, as seen in FIG. I.

Seventh operation, FIG. 8: Punch a hole and split open, as shown.

Eighth operation, FIG. 9: Place a mandril in between, and hammer down to shape as shown. Finish off by rounding the end.

Blacksmith's Manual Illustrated

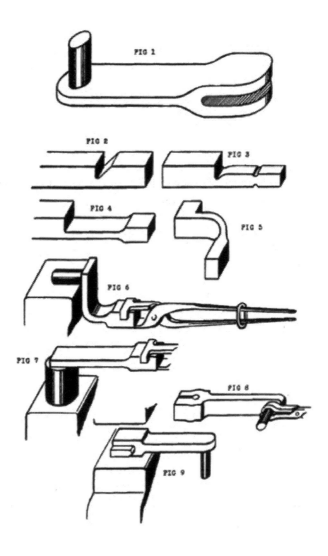

LEVER

PLATE 55: FIG. 1 illustrates a lever, made from a 2-inch diameter bar.

First operation, FIG. 2: Punch a hole 4 ins. from the end, and cut it open as shown.

Second operation, FIG. 3: Set back the ends, then cut off the bar at the dotted line as shown.

Third operation, FIG. 4: Square each end as shown.

Fourth operation, FIG. 5: Fuller close into the neck as shown.

Fifth operation, FIG. 6: Draw down to 1 in. diameter, then draw down the opposite end, and complete by bending one end, as seen in FIG. I.

The same lever can be made out of a square bar. Instead of punching a hole, and splitting as in FIG. 2, side set the bar, as shown in FIG. 7, and draw the ends down, then adopt the same operations as previously shown.

Another method of making a similar lever is to form it out of a square bar as follows: first form the boss with the radius fullers, as shown in FIG. 9, and then adopt the same methods that have been shown.

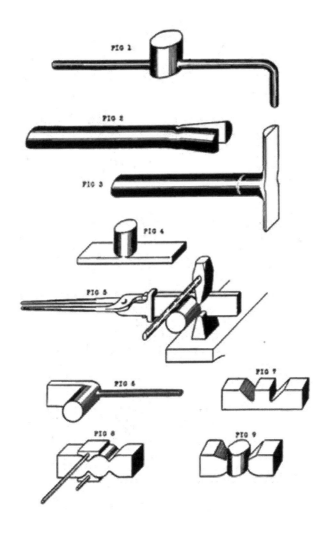

BELL CRANK LEVER

PLATE 56: FIG. 1 illustrates a bell crank lever, made from a 2-inch square bar.

First operation, FIG. 2: Bend the 2-inch square bar at right angles as shown, then cut off the bar, making both ends alike.

Second operation, FIG. 3: Fuller it as shown, leaving enough for the top and bottom boss.

Third operation, FIG. 4: Draw down each end to size, then cut the bosses to shape as shown, and finish off as required.

Another method by which the same lever can be made is as follows: Take a 5-inch by 2-inch bar and draw down, as shown in FIG. 5. FIG. 6 shows it drawn down to size. Cut off at the dotted line, and then adopt methods as previously shown.

A third method to form the bosses is illustrated in FIG. 7, by placing a pair of rings, as shown, and hammering them down. This gives the result shown in FIG. 8. Draw down, as in FIG. 9, and finish off by rounding the ends.

Note.—When using rings to shape bosses, as in FIG. 7, the material should be ¼ in. thicker than that when the chisel is used.

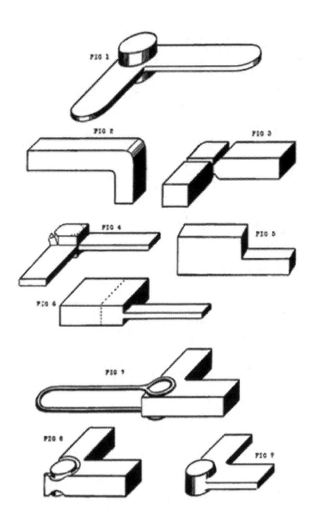

BELL CRANK LEVER

PLATE 57: FIG. 1 illustrates a bell crank lever, with the arms at right angles and the boss between them, made from a 2-inch diameter bar.

First operation, FIG. 2: Draw down 3 ins. of the 2-inch diameter bar as shown.

Second operation, FIG. 3: Joggle it by placing the 2-inch diameter bar partly over a swage, using the fuller as shown.

Third operation, FIG. 4: Draw down to size as shown.

Fourth operation, FIG. 5: Draw down the opposite end, leaving 3 ins. of the 2-inch diameter bar as shown. Next, joggle it as seen in FIG. 6, and draw down to size, bending at right angles as in FIG. 7.

Another method of making the same lever is as follows: Take a 5-inch by 2-inch bar and side set, as shown in FIG. 8. Draw down each end as in FIG. 9, fuller the inside corners, and cut around the dotted lines to form the boss as shown. When this is done, heat the boss between the two arms and twist to right angles.

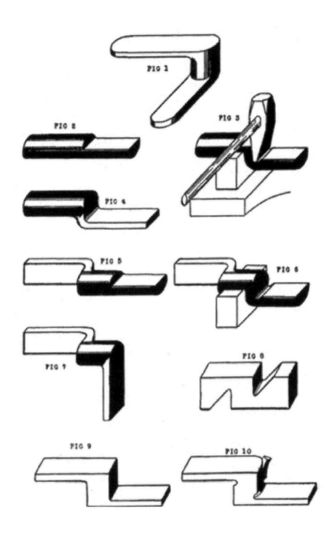

LEVER

PLATE 58: FIG. 1 illustrates a lever, made from a 3-inch square bar.

First operation, FIG. 2: Form A, using a V-shaped tool as shown.

Second operation, FIG. 3: Draw down from A to the size required to make the end boss, and side set as shown.

Third operation, FIG. 4: Draw down between the two bosses as shown.

Fourth operation, FIG. 5: Draw down the bar at the other side of A before side setting as shown.

Fifth operation, FIG. 6: Place A in a V-shaped block, side set as shown, and draw down between the side sets. The result of this operation is shown in FIG. 7 at B.

Sixth operation, FIG. 7: Fuller along the dotted line and draw down, as shown in FIG. 8.

Seventh operation, FIG. 9: Taper the forging to the required length, and complete by rounding the ends.

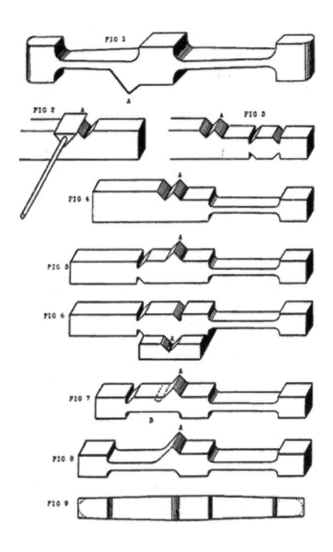

STAY

PLATE 59: FIG. 1 illustrates a stay 3 ft. 2 ins. long, made from a 6-inch by 3-inch bar.

First operation, FIG. 2: Side set as shown, making the distance between the two side sets 6 ins.

Second operation, FIG. 3: Draw down to 3 ft. and cut the ends to the required length as shown.

The same stay can be made from a 3 inch square bar 18 ins. long.

First operation, FIG. 4: Side set as shown, making the distance between the side sets 12 ins.

Second operation, FIG. 5: Draw down as shown.

Third operation, FIG. 6: Punch a hole in each end, then cut and open out as shown.

Another method of making the same stay from a 3-inch by ½-inch bar is seen in FIG. 7.

This shows a piece of 3-inch by 1-inch bar which is jumped in the centre, and fullered, as shown in FIG. 10.

FIG. 8 shows a piece of 3-inch by 1-inch bar which is jumped on the end, as shown in FIG. 9, and then welded together. Repeat the same process at the opposite end.

Blacksmith's Manual Illustrated

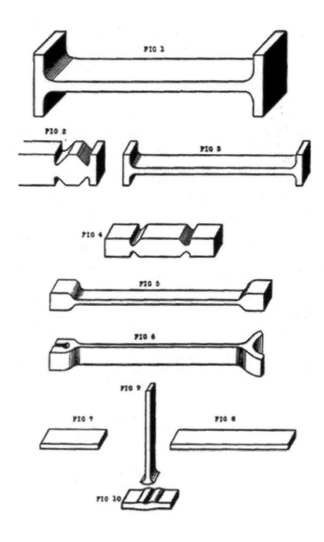

STAY

PLATE 60: FIG. 1 illustrates a stay, 22 ins. long and 1 in. square throughout, having two arms 6 ins. apart, made from a 2½-inch by 1-inch bar.

First operation, FIG. 2: Fuller 3 ins. from the end of the bar, cut along the dotted line, taking this portion out, and draw down, as shown in FIG. 3.

Second operation, FIG. 3: Side set as shown, and draw down, as in FIG. 4.

Third operation, FIG. 4: Repeat the same process as in FIG. 2 at the opposite end.

Fourth operation, FIG. 5: Drill holes as shown, cut along the dotted lines, and then open the arms out, commencing with one of the middle arms, as shown in FIG. 6. Open out the end as in FIG. 7.

FIG. 8 illustrates how, by bending the forging, the smith is enabled to dress the ends to shape.

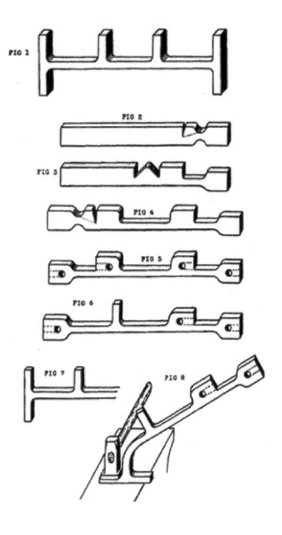

FRAME

PLATE 61: FIG. 1 illustrates a frame, which is 5 ft. by 4 ft., made from a 2-inch by ½-inch bar.

First operation, FIG. 2: Cut off two bars of the 2-inch by ½-inch bar, 5 ft. long; cut each end to an angle of 45 degrees as shown, and scarf them ready for welding.

Second operation, FIG. 3: Cut off two bars of the 2-inch by ½-inch bar, 4 ft. long, and follow the same procedure as FIG. I.

Third operation, FIG. 4: Weld the 5 ft. bar to a 4 ft. bar as shown, and repeat the same process to the other two bars as seen in FIG. 5; next weld A and B together.

Fourth operation, FIG. 6: Complete the frame by holding the two scarfs together with a pair of straps as shown.

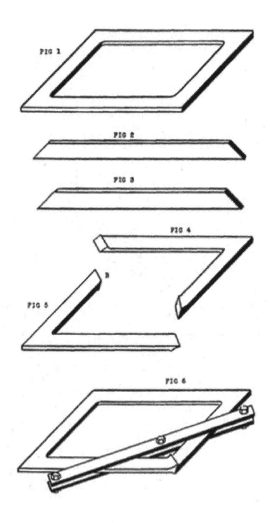

BOX SPANNER

PLATE 62: FIG. 1 illustrates a box spanner, made from a 2½-inch diameter bar, 9 ins. long.

First operation, FIG. 2: Punch a square hole in the end of the bar the required depth down.

Second operation, FIG. 3: Hammer a square mandril into the hole. This keeps the hole in shape while swaging, as shown in FIG. 4.

Note.—Mandrils used in making box spanners should have notches inserted, to enable them to be drawn out by a chisel.

Third operation, FIG. 5: Fuller as shown, and draw down as in FIG. 6.

Fourth operation, FIG. 7: Flatten the end to form the T-piece, punch a hole, and split open as shown.

Fifth operation, FIG. 8: Bend as shown. This enables the smith to draw down the ends as shown.

Another method of making a box spanner is as follows :—

FIG. 9: Make a collar out of 2-inch by ¾-inch bar as shown, and place on the end of a 1-inch diameter bar, as in FIG. 10. Weld the two together, and when this is done, drill a hole into the collar and square it by hammering the mandril in, as shown in FIG. 3.

FIG. 11 shows a method of making a T-piece, by welding two round bars together.

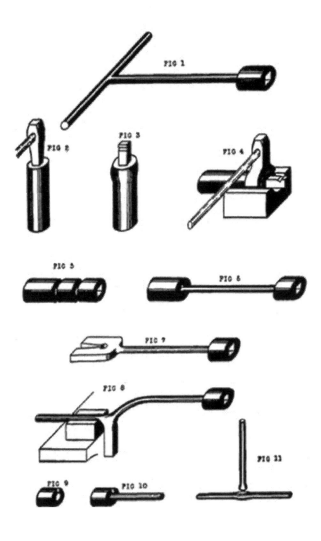

HINGES

PLATE 63: FIG. 1 illustrates a pair of hinges, made from a 4-inch by 2-inch bar.

First operation, FIG. 2: Fuller the bar 2 ins. from the end as shown, and draw down as in FIG. 3. Cut off from the bar along dotted line, cut the corners off, and finish off the boss.

Second operation, FIG. 4: Punch or drill a hole as shown, and cut the centre portion out along the dotted lines. With a little dressing up this will complete the double hinge.

Third operation, FIG. 5: Make a forging similar to above, and cut the outside portions off along the dotted lines as shown.

In FIG. 6 another method of forming the boss of a hinge, by stamping into a swage, is shown. FIG. 7 shows the result. This only needs one corner cut off.

Another method of making hinges is from a flat bar as follows:—

First operation, FIG. 8: Point the end, and 2 ins. from it cut through the bar as shown.

Second operation, FIG. 9: Bend it to shape on a mandril as shown.

Note.—FIG. 9 shows the hinge welded on the top. It can be welded underneath, the latter method being preferable.

Third operation, FIG. 10: Flatten the two outside eyes down as shown, and cut them off, leaving the single hinge. To make the double hinge, flatten the centre portion (FIG. 11) and cut it off.

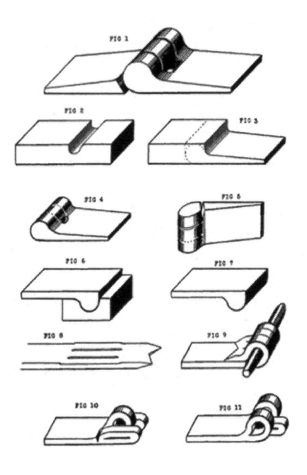

CLAMS

PLATE 64: FIG. 1 illustrates a pair of clams, made from 3-inch by 2-inch bar, showing the corners thicker than the rest of the clams.

First operation, FIG. 2: Fuller the bar as shown.

Second operation, FIG. 3: Draw down the ends and centre of the bar, leaving two points. These form the inside corners of the clams.

Third operation, FIG. 4: Bend, in a V-block, the points, as shown in FIG. 5.

Fourth operation, FIG. 6: Shape the bar as shown. This simplifies the bending of the other end.

Another method of forming the corners is by nicking a flat bar, as shown in FIG. 7, and bending it at right angles (FIG. 8). A round bar is then welded across the corner, as in FIG. 9. Finish off as previously stated.

Blacksmith's Manual Illustrated

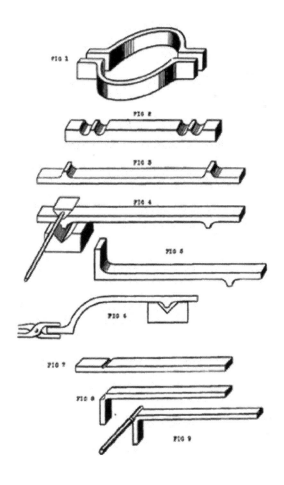

AXE

PLATE 65: FIG. 1 illustrates an axe, made from 2-inch by ½-inch iron or mild steel bar.

First operation, FIG. 2: Weld a piece of blister steel, 1 in. by ½ in., 4 ins. from the end of the bar, as shown.

Second operation, FIG. 3: Spread out with a fuller on each side of the blister steel, as shown.

Third operation, FIG. 4: Double over as shown, and weld the two ends together.

Fourth operation. FIG. 5: Prepare a piece of blister steel, wedge shape as shown. Split the end of the forging, and place the blister steel in between, as shown in FIG. 6. Raise to a welding heat and flatten out to form the blade.

When hardening axes, made as described above, heat to a dark red, and after plunging into water, cover with oil. Heat over the fire until the oil ignites, and finish by cooling off.

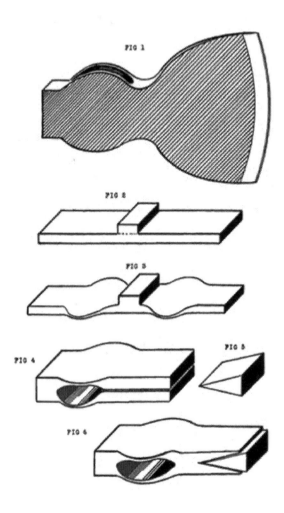

AXE

PLATE 66: FIG. 1 illustrates another type of axe, generally used in and about coal mines, made from 1½-inch square iron or mild steel bar.

First operation, FIG. 2: Prepare a piece of blister steel 1 ½ in. square by ¼ in. thick as shown, and a wedge-shaped piece, as seen in FIG. 3.

Second operation, FIG. 4: Fix the blister steel (FIG. I) on the end of the 1½-inch square bar as shown, and weld them together.

Third operation, FIG. 5: Fuller the corners of the bar, 1 in. from the end, as shown.

Fourth operation, FIG. 6: Punch a hole as shown, hammer a mandril in, and draw out each side of the eye while the mandril is still in, using a fuller, as shown in FIG. 7. When this is done, replace the former mandril with a larger or finishing one, and cut the required length from the bar.

Fifth operation, FIG. 8: Split the end as shown, and place the blister steel (FIG. 3) in between. Raise to a welding heat, and flatten out to form the blade.

This axe is hardened by the same method as the axe on PLATE 65.

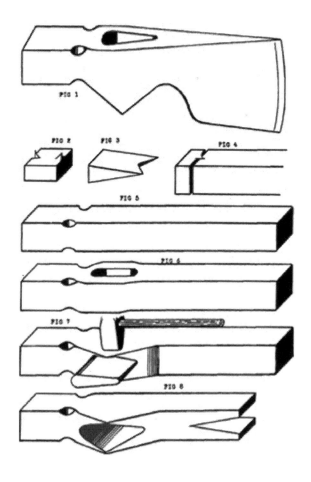

THUMB SCREWS

PLATE 67: FIG. 1 illustrates a thumb screw, made from 1¼-inch square bar.

First operation, FIG. 2: Place the 1¼-inch square bar into a pair of 1-inch swages corner-wise as shown, using the swages in this case as stamps; result is shown in FIG. 3. Next cut along the dotted line, leaving the result as shown in FIG. 4.

Second operation, FIG. 5: Draw down to size as shown.

Third operation, FIG. 6: Hammer two narrow blocks down as shown, then cut from the bar, leaving the result as seen in FIG. 7.

Fourth operation, FIG. 8: Cut out along the dotted lines, leaving the result as shown in FIG. 9. Complete by cutting the corners off.

FIG. 10 shows another thumb screw, made from a round bar, as follows :—

First operation, FIG. 11: Draw down to the required diameter, fuller as shown, then cut to length from the bar, and complete by flattening to size.

Special spring swages can be made to form the ball and collar at the same time.

FIG. 12 shows the result after using these swages.

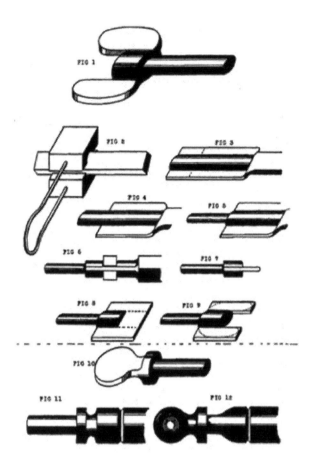

TUB CROOK

PLATE 68: FIG. 1 illustrates a tub crook, made from 2½-inch square bar.

First operation, FIG. 2: Side set the bar and draw down, as shown.

Second operation, FIG. 3: Fuller, using two small bars as shown, then flatten down, as seen in FIGS. 4 and 5.

Third operation, FIG. 6: Side set as shown, and draw down, as in FIG. 7.

Fourth operation, FIG. 8: Bend down as shown, and cut to shape at the dotted lines.

Fifth operation, FIG. 9: Hammer into shape, then cut and swage the corners on the outside and inside of the crook along the dotted lines, as shown.

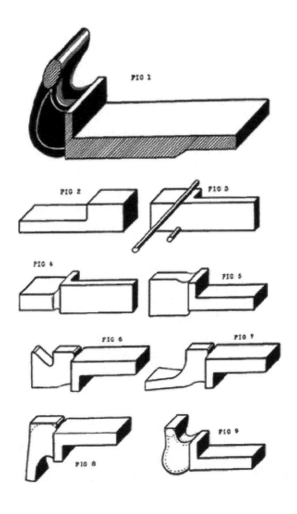

HOOKS

PLATE 69: FIG. 1 illustrates a hook, made from 1½-inch diameter bar. A very useful rule to follow, when the diameter of the hook is given, is as follows :—

Total length of hook from the neck to the end to form bend should be diameter multiplied by 8, e.g. a 2-inch diameter hook would require 16 ins. of material from neck to point.

First operation, FIG. 2: Form a ball on the end of the 1-inch diameter bar. This is done by using ball swages, as shown in FIG. 8.

Second operation, FIG. 3: Flatten the ball as shown, and punch a hole.

Third operation, FIG. 4: Set the neck of the hook through, using a radius fuller, as shown.

Fourth operation, FIG. 5: Taper the end to a point, and bend it over the beak of the anvil, as shown.

Fifth operation, FIG. 6: Bend to shape, by placing the fuller on the end, which is kept cool. Finish by hammering on top while some one is holding on with a hammer, as shown.

FIG. 7 shows how, by careful handling, a hook can be bent to shape under the hammer.

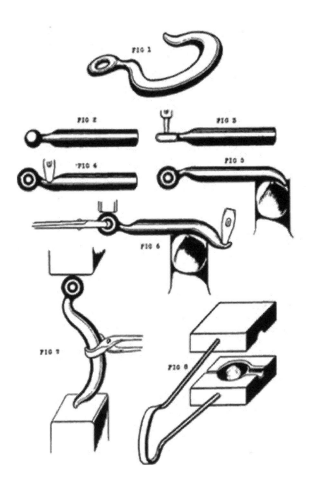

DOUBLE "S" LINK

PLATE 70: FIG. 1 illustrates a double "S" link which can be used, in an emergency, to connect a broken chain. This can be made from any small size diameter bar.

First operation, FIG. 2: Punch a hole in the centre of the bar, and draw each end down, as shown.

Second operation, FIG. 3: Bend the bar edgeways around a template, as shown. By using this template, all the ends can be made alike.

FIG. 4 shows the shape of half the double "S" link before the parts are riveted together, as shown in FIG. 1.

FIG. 5 shows the link in use, connecting a broken chain.

Blacksmith's Manual Illustrated

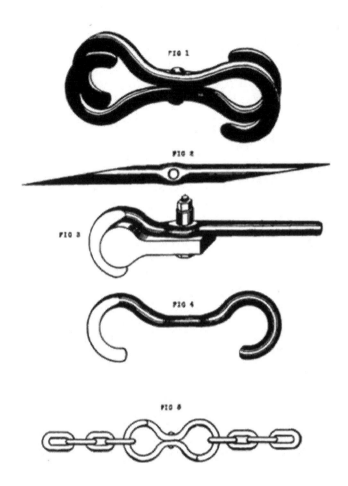

SOCKET

PLATE 71 illustrates a socket, used for haulage ropes. These are made in different sizes to suit the various sizes of ropes. FIG. I, as an example, is made from 1½-inch diameter bar, 6 ins. long.

First operation, FIG. 2: Fuller 2 ins. from the end of the bar, and draw down, as shown in FIG. 3.

Second operation, FIG. 4: Punch or drill two holes as shown, then cut out between the two holes, as shown in FIG. 5.

Third operation, FIG. 6: Taper the 1½ in. diameter down to 1¼ in. diameter as shown, then drill a hole through, as shown in FIG. 7.

Fourth operation, FIG. 8: Bend over as shown, to enable a tapered hand mandril to be hammered in, as shown in FIG. 9, then straighten to shape again.

FIG. 10 shows the method of socketing, by putting the rope through. A wire is then lapped around 3 ins. from the end, and a $\frac{3}{16}$-inch split ring fixed on, as shown in FIG. 11. Complete by opening the end of the rope out, turning over the ends, and bending together by overlapping with wire, as shown in FIG. 12.

Blacksmith's Manual Illustrated

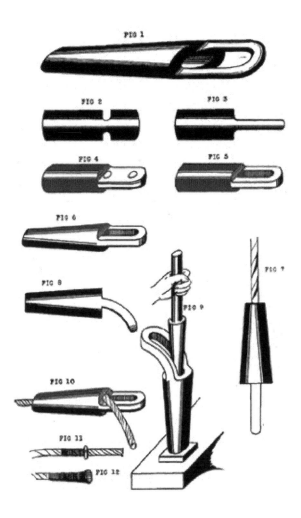

KNOCK OFF

PLATE 72: FIG. 1 illustrates a knock off, generally used in coal mines. These are fixed to the coal tubs, as shown in FIG. 7, and automatically release the tubs from the haulage rope. At the point where the tubs have to be released, a bar is fixed above (FIG. 8) which knocks back the lever and releases the pin at the same time. The method of making a knock off is as follows :—

First operation, FIG. 2: Side set a 2½-inch square bar as shown, then draw down, as in FIG. 3.

Second operation, FIG. 4: Punch or drill a hole as shown, and cut out at the dotted lines.

Third operation, FIG. 5: Punch a hole and split open, insert a flat mandril, as shown in FIG. 6, and flatten down. To complete, as illustrated in FIG. 1, bend a 2-inch by $\frac{5}{8}$-inch flat bar at right angles, making one end 10 ins. and the other 8 ins., then rivet together, as shown.

Blacksmith's Manual Illustrated

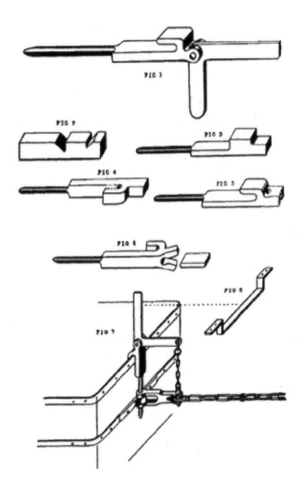

DRILL STAND

PLATE 73: FIG. 1 illustrates a drill stand, showing the lever arm. This arm can be altered to any convenient height or direction.

First operation, FIG. 2: Side set a 3-inch by 1½-inch bar as shown, then draw down, as shown in FIG. 3.

Second operation, FIG. 4: Bend the ends around as shown, and complete the forked end by placing a mandril in, and hammering down, as shown in FIG. 5.

Third operation, FIG. 6: Jump a 1½ in. diameter as shown, and weld together, as in FIG. 7.

Another method of making the forked end is shown in FIG. 8. Bend two flat bars and weld them together.

FIG. 9 shows another method by drawing down two 1½-inch square bars and welding them together.

FIG. 10 shows a piece of 3-inch by 1½-inch bar drawn down, the radius cutters having been used to form the end.

FIG. 11 shows the hole drilled, and a key-way cut through.

FIG. 12 shows a method of making the key by drawing down the end of the bar and turning the end up.

Blacksmith's Manual Illustrated

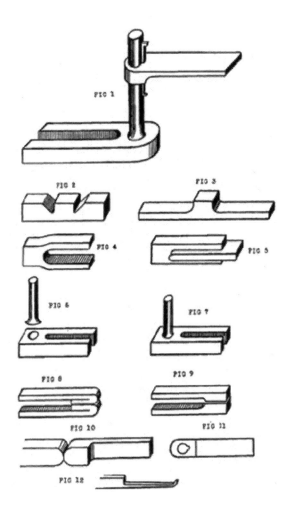

FORGING

PLATE 74: FIG. 1 illustrates a forging, made from a 6-inch square bar. The dimensions of this forging are given herewith :—

A is 10 ins. by 6 ins. and 1¼ in. thick.

B is 2¾ ins. diameter and 4 ins. high.

C is 4 ins. by 2¾ ins. and 4 ins. deep.

First operation, FIG. 2: Side set the bar as shown, then draw down to 2¾ ins. diameter, as in FIG. 3. Cut off from the bar at the dotted lines, 6 ins. long.

Second operation, FIG. 4: Place in a bolster, and side set as shown, then hammer down, as in FIG. 5, to 1¼ in. thick.

Third operation. FIG. 6: Use two $\frac{5}{8}$-inch bars and fuller as shown, then draw down as in FIG. 7, and complete by cutting to length.

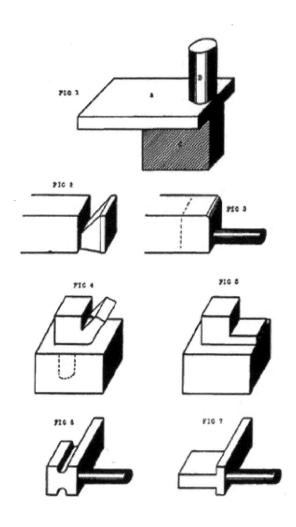

FORGING

PLATE 75 illustrates a rather complicated forging made from a 6-inch square bar. The dimensions of this forging are here given :—

A 3¼ ins. square.

B 2 ins. diameter, 4 ins. long.

C 3½ ins. by ⅝ in. by 7 ins. long, and the height from the bottom of C to the top of A is 6 ins.

First operation, FIG. 2: Side set two sides of the 6-inch square bar as shown, and draw down to 3¼ ins. square, as in FIG. 3.

Second operation, FIG. 4: Side set the 3¼ ins. square on each side as shown, then draw down to 2 ins. diameter, as in FIG. 5. Next cut it off the bar at the dotted line, 3¼ ins. long.

Third operation, FIG. 6: Side set as shown, and draw down, as in FIG. 7, then cut along the dotted line, leaving the result, as shown in FIG. 8.

Fourth operation, FIG. 9: Fuller as shown, and cut the two portions off along the dotted line, or flatten them down and cut to shape, as shown in FIG. 10.

Blacksmith's Manual Illustrated

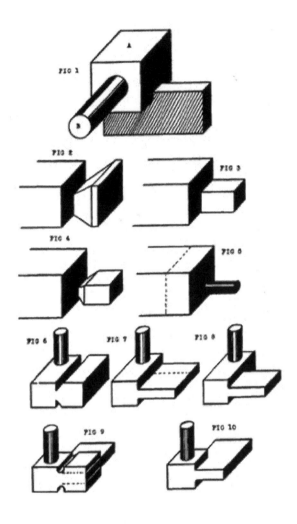

LEVELLING BLOCK

Levelling blocks are very useful in the smithy, their dimensions depending on the size of work chiefly done. For general use, a block, as illustrated in FIG. 1, measures 6 ft. by 4 ft. by 4 ins. thick, numerous holes being left in the casting.

FIG. 1 illustrates a method of bending large bars, by placing tapered pins in holes where necessary. By making use of these together with a lever, the bar is bent, as shown.

FIG. 2 illustrates a method of bending large links. The required size of the link can be obtained by selecting a bolster, placing it over a pin and pulling the bar around it, as shown. A method of clamping plates or bars to be levelled is also shown.

FIG. 3 illustrates a cramp or rail bender, used for bending rails. It can also be used for bending heavy angle bars.

Blacksmith's Manual Illustrated

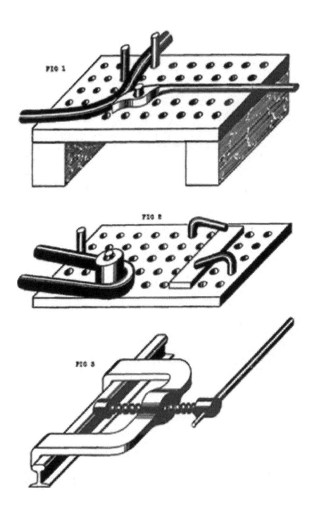

STAMPS

PLATE 77: FIG. 1 illustrates a half stamp. A method of making such a stamp is given in the following illustrations:—

First operation, FIG. 2: Fuller a 6-inch square block 2 ins. thick, as shown. Next, hammer a small diameter bar into the block (FIG. 3), and gradually increase the diameter until the required size is obtained (FIG. 4).

Second operation, FIG. 5: Place a flat bar on the stamp as shown, and hammer it in, giving the result shown in FIG. 6.

Third operation, FIG. 7: Place a smaller but thicker bar on the stamp as shown, and hammer it in, giving the result shown in FIG. 1.

Repeat these operations on a similar block, thus making a pair of stamps.

By placing a finished forging (FIG. 8) in the stamps (FIG. 9) a correct finish can be obtained.

Pins (FIG. 10) are screwed into the bottom stamp, as shown in FIG. 11, to enable the stamps to be exactly above one another during operations.

Note.—Stamps are generally used when large quantities of the same job are required.

Blacksmith's Manual Illustrated

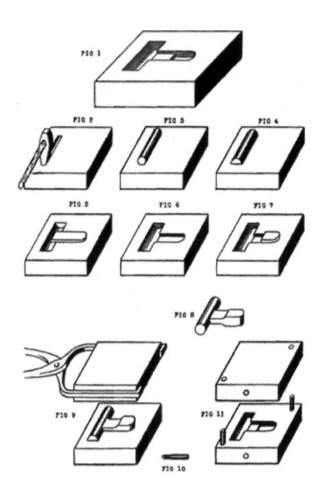

STAMPS

PLATE 78: FIG. 1 illustrates a finished forging made under the steam hammer, with stamps.

FIG. 2 illustrates a pair of stamps complete with handles. The method of making these stamps was shown on PLATE 77.

FIG. 3 shows the forging after being stamped.

FIG. 4 shows a ragging tool, used under the steam hammer, for taking away the ragged edges from the forging after it has been stamped.

FIG. 5 shows the ragging tool in use.

Method.—Place the unfinished forging on the ragging tool, and place a shaped bar on top, as shown. Strike with the steam hammer, pressing the finished forging through, and leaving the ragged edges behind.

FIGS. 6 to 8 show the preparing of the forging before it is placed in the stamps.

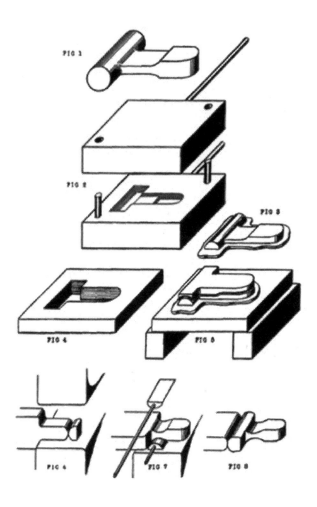

SPANNER STAMPS

PLATE 79: FIG. 1 illustrates a double-ended spanner, which can be made in the stamps shown in the following illustrations:—

FIG. 2 shows a pair of stamps used for shaping the material after it has been prepared, as in FIG. 5.

FIG. 3 shows a punch used for punching the jaw in the spanners, while they are in FIG. 4 stamps.

FIG. 4 shows a pair of stamps used for finishing the spanners after they have been shaped in FIG. 2.

FIG. 5 shows a round bar, balled at each end ready to be shaped by placing it in FIG. 2 stamps.

FIG. 6 shows another pair of stamps used for shaping large spanners.

FIG. 7 shows FIG. 6 stamps in use.

FIG. 8 shows the material roughly shaped before placing in the stamps.

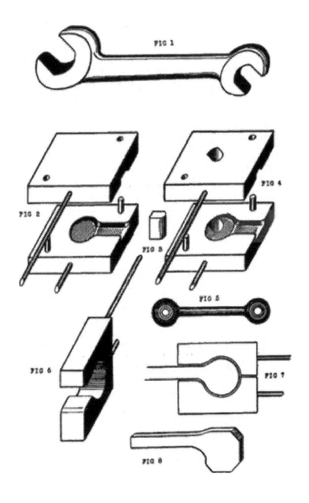

LOCOMOTIVE FLY CRANK

PLATE 80: FIG. 1 illustrates a fly crank, made from a 5-inch square bar.

First operation, FIG. 2: Draw down the 5-inch square bar, as shown.

Second operation, FIG. 3: Side set, as shown.

Third operation, FIG. 4: Draw down, and fuller at B, then draw down as shown at A, and cut off at the dotted line.

Note.—These illustrations give a method of forging two fly cranks at once.

Fourth operation, FIG. 5: After cutting along the dotted line, bend the forging as shown, in order to swage the pin.

Fifth operation, FIG. 6: Straighten the forging, then place it in a bolster and flatten to size. Finish off by shaping, as shown in FIG. 1.

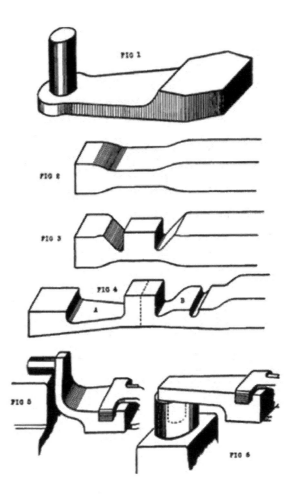

LOCOMOTIVE ECCENTRIC ROD

PLATE 81: FIG. 1 illustrates an eccentric rod, made from a 6-inch square bar. The dimensions are :—

A 8 ins. by 3 ins. by 1½ in.

B 3 ft. 5 ins. long and 1 in. thick, tapering from 3¾ ins. to 2¼ ins.

C 2¾ ins. diameter, 5 ins. deep.

The length of material required to make this is 13 ins.

First operation, FIG. 2: Side set 11½ ins. from the end of the bar.

Second operation, FIG. 3: Draw down from 6 ins. square to 5 ins. by 3 ins.

Third operation, FIG. 4: Side set 15 ins. from the 6-inch square bar, and side set again, as shown.

Fourth operation, FIG. 5: Draw down to length as shown, and shape the double eye end, using radius cutters.

Fifth operation, FIG. 6: Finish the double eye end by stamping in swages, as shown. Complete by cutting the opposite end off the bar and hammering down to 8 ins. by 3 ins. by 1½ in.

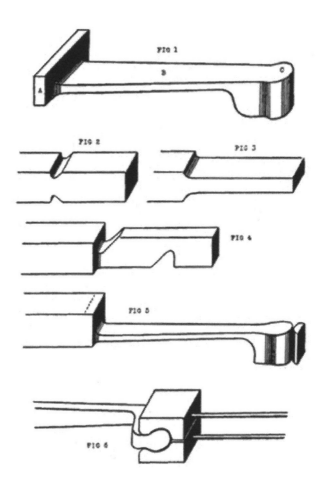

LOCOMOTIVE CROSS BEAM

PLATE 82: FIG. 1 illustrates a cross beam, made from a 4-inch square bar.

First operation, FIG. 2: Fuller the 4-inch square bar, as shown.

Second operation, FIG. 3: Draw each end down and shape, as shown.

Third operation, FIG. 4: Draw the bar down and flatten out to the required width, as shown in FIGS. 4 and 5.

Fourth operation, FIG. 6: Taper the forging from the centre as shown, side set the ends, adopting the method shown in FIG. 7, and draw down the ends to the required diameter (FIG. 8).

Blacksmith's Manual Illustrated

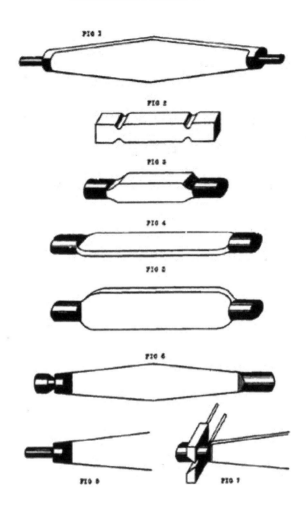

LOCOMOTIVE REVERSING LEVER

PLATE 83: FIG. 1 illustrates a reversing lever, made from a 3-incb square bar.

First operation, FIG. 2: Side set the 3-inch square bar as shown, adopting the method given in FIG. 8.

Second operation, FIG. 3: Set the bar through, then side set, as shown.

Third operation, FIG. 4: Again set the bar through, and side set, as shown.

Fourth operation, FIG. 5: Partly draw down, and side set, as shown.

Fifth operation, FIG. 6: Draw the bar down, and form the boss, as shown. Next, side set the opposite end as shown.

Sixth operation, FIG. 7: Draw down as shown, then finish off the end, as in FIG. 1.

FIG. 8 shows the method of side setting used in the first and second operations. First, side set at one side, then turn over the bar, and complete by using two side set tools, as shown.

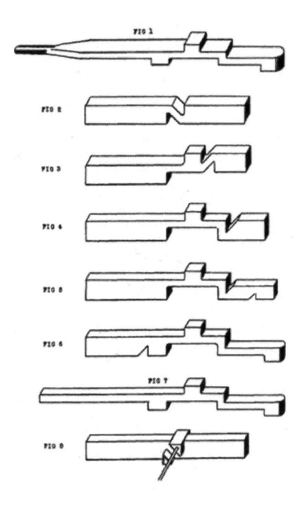

LOCOMOTIVE BRAKE HANGER

PLATE 84: FIG. 1 illustrates a brake hanger, made from a 4-inch square bar.

First operation, FIG. 2: Side set the 4-inch square bar, as shown.

Second operation, FIG. 3: Draw down and flatten out to the width required, as shown.

Third operation, FIG. 4: Side set, as shown.

Fourth operation, FIG. 5: Draw down to the thickness of the centre boss, and fuller as shown.

Fifth operation, FIG. 6: Draw down to thickness in between the bosses as shown, and then complete the forging by cutting to shape on the dotted lines, as shown.

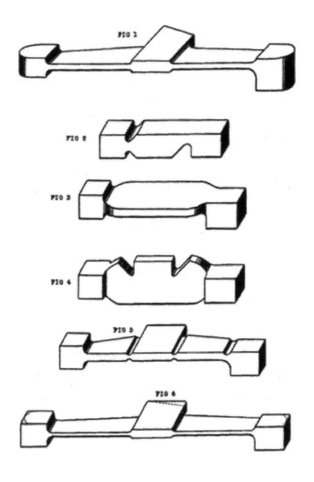

LOCOMOTIVE BRAKE HANGER

PLATE 85: FIG. 1 illustrates a brake hanger, made from 3½-inch square bar.

First operation, FIG. 2: Side set the bar, as shown.

Second operation, FIG. 3: Draw down to the required thickness. Repeat the above operations, at the other end except that the end to be side set is opposite.

Third operation, FIG. 4: After drawing down as shown, complete by cutting the bosses to shape.

FIG. 5 shows how, by fullering the inside corners, the work is simplified. It is easier to cut the corners off, and this method also prevents the chisel from cutting the bar.

FIG. 6 shows the boss being shaped, after the inside corners have been fullered.

FIG. 7 shows another method of forming the bosses by stamping a pair of loose rings into the bar, as shown.

FIG. 8 shows a split swage on which bosses can be rounded, as shown.

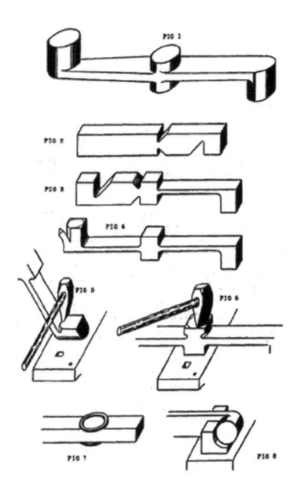

LOCOMOTIVE BRAKE HANGERS

PLATE 86: FIG. 1 illustrates a brake hanger, made from a 2½-inch square bar. The total length required is 10¼ ins. The dimensions are :—

A 2½ins. diameter by 2¼ins. thick.

B The greatest width is 3 ins. tapered down to 1¾ in. at each end, and 1 in. thick.

First operation, FIG. 2: Use radius cutters to shape the end of the boss as shown, leaving enough material to make a tong end, as shown in FIG. 4. Next, draw down 5¼ ins. of the square bar after fullering, as shown in FIGS. 2 and 4.

FIG. 5 shows the method of tapering B, by using tapered sets illustrated in FIG. 7.

FIG. 6 illustrates a pair of stamping swages used to form the bosses A.

BRAKE HANGERS.

PLATE 86

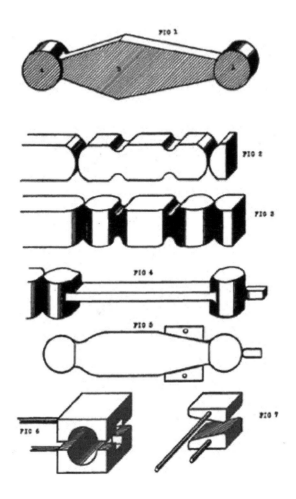

LOCOMOTIVE CONNECTING ROD STRAP

PLATE 87: FIG. 1 illustrates a connecting rod strap, made from 4½-inch square bar.

First operation, FIG. 2: Side set the 4½-inch square bar, as shown.

Second operation, FIG. 3: Partly draw down each side of the studs, as shown.

Third operation, FIG. 4: Draw down to the smallest size, as shown.

The reason for forming two studs when forging, is to help when bending, by placing a bridge piece, which is made to the required size, on the forging, as shown in FIG. 5; hammer down, as shown in FIG. 6, then cut off the stud which is not required.

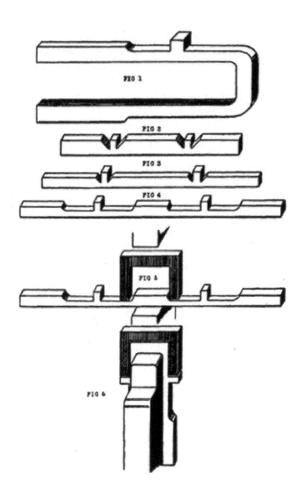

LOCOMOTIVE HORN STAY

PLATE 88: FIG. 1 illustrates a horn stay, made from 6-inch by 4-inch bar.

First operation, FIG. 2: Side set the 6-inch by 4-inch bar, as shown.

Second operation, FIG. 3: Draw down to size, side set the ends, and draw down, as shown. When drawing down in between the studs use a narrow block.

Third operation, FIG. 4: Use a pair of stamps and hammer down, as shown.

Fourth operation, FIG. 5: Shows the result after using the stamps.

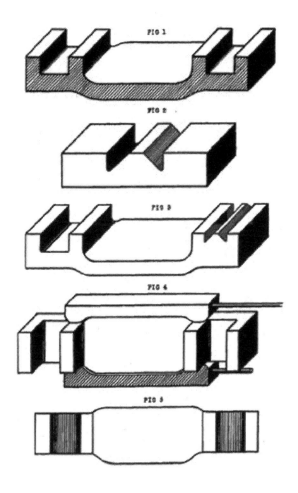

LOCOMOTIVE BRAKE SHAFT (A)

PLATE 89: FIG. 1 illustrates a brake shaft, 4 ft. 6 ins. long, made from 9-inch square bar.

First operation, FIG. 2: Side set the 9-inch square bar as shown, then draw down to 9 ins. by 4 ins., as shown in FIGS. 3 and 4.

Third operation, FIG. 5: Side set as shown, draw down to 3½ ins. diameter, and cut half of the short arms off where marked. The result of these operations are shown in FIG. 6.

Fourth operation, FIG. 6: Drill holes as shown. The forging is then taken to a band saw to be cut along the dotted lines, leaving the result as shown in FIG. 7.

Fifth operation, FIG. 8: Open the arms out, and taper them to the required size. Swage the rest of the shaft to 3½ ins. diameter, as shown.

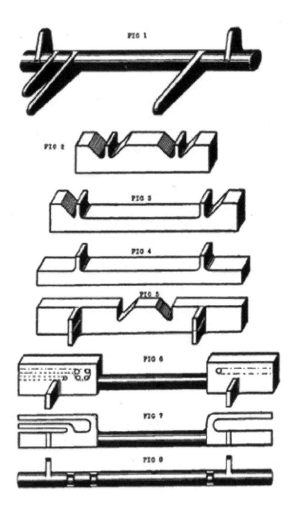

LOCOMOTIVE BRAKE SHAFT (B)

PLATE 90 illustrates the method of making a brake shaft having an arm with a double eye at one end, and a straight arm at the other end.

FIG. 1 shows the brake shaft. The dimensions are :—

A 3¼ ins. diameter, 2 ft. 4 ins. long.

B 4 ins. tapered to 3¼ ins., 12 ins. long and 1½ in. thick.

C the largest diameter is 4 ins., 4 ins. thick. The arm is 1½ in. thick, total length 12 ins. and the distance between the two arms is 1 ft. 4 ins. The size of bar required to make the brake shaft is 8 ins. by 4 ins.

First operation, FIG. 2: Draw down 15 ins. of the 8-inch by 4-inch bar to 6 ins. by 4 ins., then side set, allowing 4½ ins. to be drawn down as shown to 3¼ ins. diameter, 1 ft. 4 ins. long.

Second operation, FIG. 3. Draw down to 3¼ ins. diameter, then cut from the bar, making each end 12 ins. long, one end being 8 ins. by 4 ins. for 6 ins. of its length.

Third operation, FIG. 4: Mark off for drilling as shown, then cut along the dotted lines.

Fourth operation, FIG. 5: Open out the arms, and shape them as required. Next, finish off the ends of the shaft and cut to length. To complete the forging, twist the bar in the centre (at the double line). This gives the required result, as shown in FIG. 1.

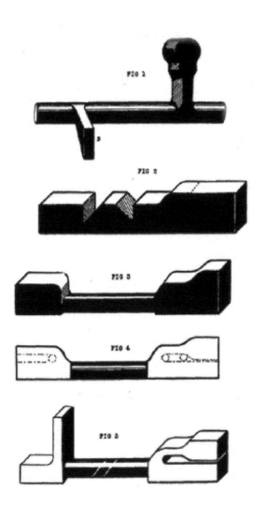

LOCOMOTIVE BRAKE SHAFT (C)

PLATE 91 illustrates the method of making a brake shaft, having a long arm in the centre, with a double eye at the end. Two short arms project from each end of the shaft, as shown.

FIG. 1 shows the brake shaft, made from an 8-inch square bar.

First operation, FIG. 2: Side set the **8**-inch square bar, as shown.

Second operation, FIG. 3: Draw down to 8 ins. by 4 ins., as shown.

Third operation, FIG. 4: Mark off the forging for drilling and cutting, as shown.

FIG. 5 shows the result after the forging has been cut with a band saw along the dotted lines.

Open out the centre arm, and then cut the piece off at the dotted line. Complete, by swaging the bar to the required diameter, and shape the double eye, as shown in FIG. 1.

Blacksmith's Manual Illustrated

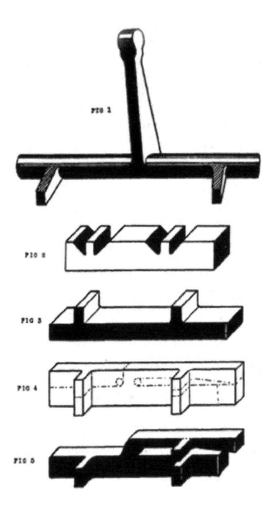

FIG 1

FIG 2

FIG 3

FIG 4

FIG 5

LOCOMOTIVE SHAFT

PLATE 92 illustrates the method of making a shaft having a short arm at each end.

FIG. 1 shows the shaft, made from a 12-inch by 6-inch bar.

First operation, FIG. 2: Side set the 12-inch by 6-inch bar, as shown.

Second operation, FIG. 3: Draw down one end and swage it to the required diameter, as shown.

Third operation, FIG. 4: Repeat the above operations at the other end.

Fourth operation, FIG. 5: Draw down and swage the centre, as shown. Next, shape the arms by cutting along the dotted lines, as shown.

FIG. 6 shows the method of side setting, by using a pair of side sets.

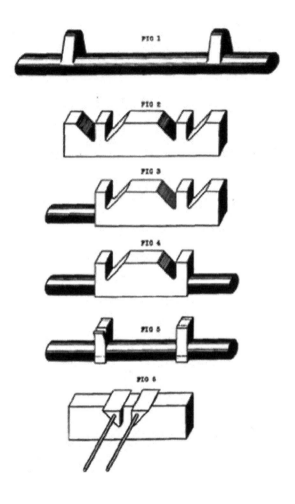

LOCOMOTIVE REVERSING SHAFT (A)

PLATE 93 illustrates the method of making part of a reversing shaft which is 17 ins. long and the diameter of which is 3¼ ins. The arm with the double eye forged on the end is 9 ins. long, and the T-piece at the opposite end is 8 iris, by 4 ins. tapered to 2 ins. at the ends.

FIG. 1 illustrates the reversing shaft made from a 7-inch square bar.

First operation, FIG. 2: Draw down 7 ins. of the 7-inch square bar to 4½ in. square. Reduce enough of the 4½-inch square to form the arm, as shown in the following illustrations.

Second operation, FIG. 3: Joggle as shown.

Third operation, FIG. 4: Swage to the required diameter and cut off the bar at the dotted line. Prepare the double eye for stamping, as shown in FIG. 5.

FIG. 6 shows the result after stamping complete by setting the arm and tapering the opposite end, as shown in FIG. 1.

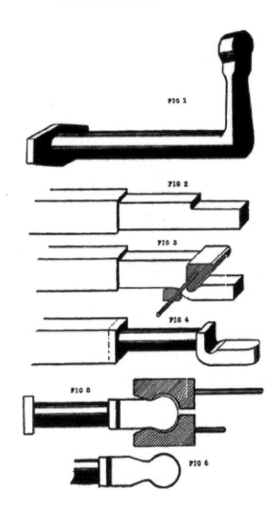

LOCOMOTIVE REVERSING SHAFT (B)

PLATE 94 illustrates the making of another part of a reversing shaft, having an extra arm in the centre.

FIG. 1 illustrates the reversing shaft, made from a 7-inch square bar.

First operation, FIG. 2: Draw down the 7-inch square bar to 7 ins. by 4 ins., as shown.

Second operation, FIG. 3: Side set the 7 ins. by 4 ins., as shown.

Third operation, FIG. 4: Draw down and swage, then fuller, as shown.

Fourth operation, FIG. 5: Reduce as shown, to shape partly the centre arm, then drill two holes and cut the bar as shown.

Fifth operation, FIG. 6: Open out and set the centre arm.

Sixth operation, FIG. 7: Turn over the bar and draw down, as shown. Next, joggle as shown in FIG. 8.

Seventh operation, FIG. 9: Draw down and swage the rest of the shaft, fuller as shown, then draw down as shown in FIG. 10. Complete the double eyes by stamping, as shown on PLATE 93, and set the end arm, as shown in FIG. 1.

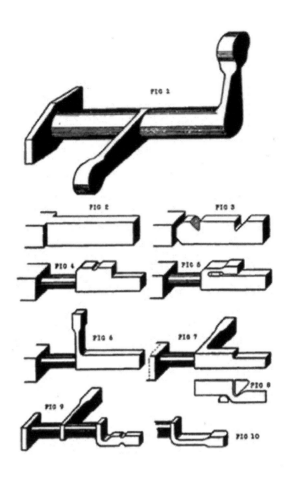

LOCOMOTIVE REVERSING SHAFT (C)

PLATE 95: FIG. 1 illustrates a reversing shaft, made from 9-inch by 4-inch bar.

First operation, FIG. 2: Side set the 9-inch by 4-inch bar, as shown.

Second operation, FIG. 3: Draw down to 6½ ins. by 4 ins. and side set, as shown.

Third operation, FIG. 4: Swage to the required diameter, then side set the opposite end, and drawdown, as shown in FIG. 5.

Fourth operation, FIG. 6: Mark off the forging for drilling and cutting, as shown.

Fifth operation, FIG. 7: Shows the result after the forging has been cut with a band saw along the dotted lines.

Sixth operation, FIG. 8: Open out the arms, and draw down, as shown.

Seventh operation, FIG. 9: Joggle as shown.

Eighth operation: Swage the remainder of the shaft to the required diameter, before setting the end arm, as shown in FIG. 1.

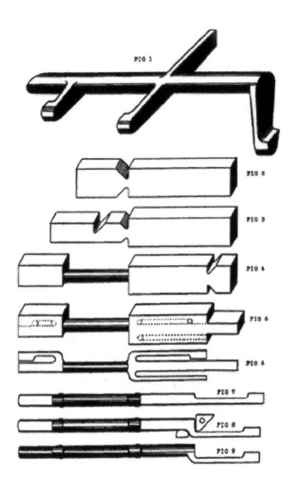

LOCOMOTIVE LOCKING BAR (A)

PLATE 96 illustrates the method of making a locking bar, which is 5 ft. 10 ins. long.

FIG. 1 illustrates the locking bar, made from a 5-inch square bar.

First operation, FIG. 2: Punch a hole in the centre of the bar, as shown.

Second operation, FIG. 3: Fuller each side of the hole, with fullers made, as shown.

Third operation, FIG. 4: Hammer in a mandril, and stamp the eye with a pair of tools, as shown.

Fourth operation, FIG. 5: Draw down and taper the ends.

FIG. 6 shows the mandril for doing the preliminary work. This is called the starting mandril.

FIG. 7 shows the mandril for finishing the shape of the hole, known as the finishing mandril.

Blacksmith's Manual Illustrated

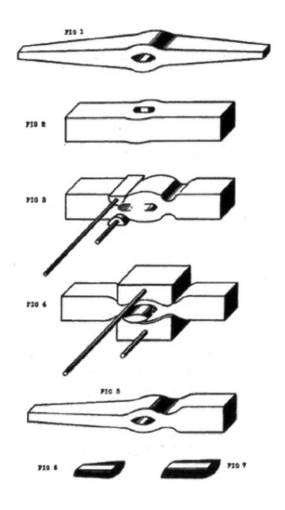

LOCOMOTIVE LOCKING BAR (B)

PLATE 97 illustrates the beginning of the method of making a locking bar.

FIG. 1 illustrates a locking bar, of which the dimensions are given, made from a 12-inch by 6-inch bar. The length of material required is 15 ins.

First operation, FIG. 2: Side set 5 ins. from the end and 10 ins. from the end, as shown.

Second operation, FIG. 3: Draw down and taper one end as shown, then cut off the bar at the dotted line.

Third operation, FIG. 4: Draw down the opposite end, as shown.

This method is continued on the following PLATE (98).

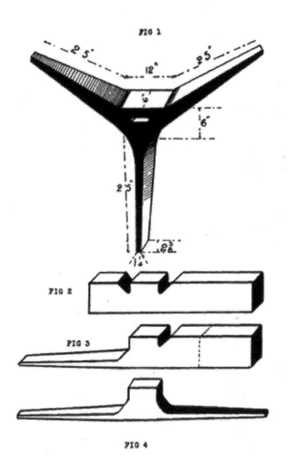

LOCOMOTIVE LOCKING BAR (B) (*continued*)

PLATE 98 illustrates the concluding operations in the method of making a locking bar.

FIG. 5 illustrates an apparatus used for setting the arms at the required angle. This apparatus is composed of two blocks of wood placed at each side of the anvil (steam hammer), and held together, as shown.

FIG. 6 shows the result after the arms have been set with the steam hammer.

FIG. 7, fuller as shown, then draw down and taper to the length, as shown in FIG. 1 (PLATE 97).

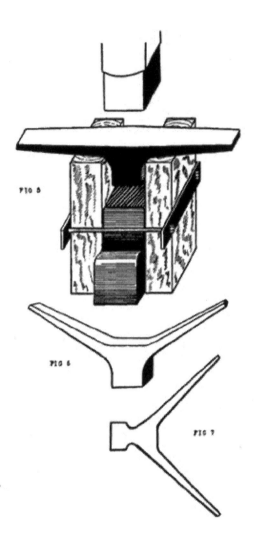

HARDENING AND TEMPERING

Small Coil Spring

FIG. 1 illustrates a method of hardening a small coil spring. This is done by placing a small rod through the spring as shown, and heating it over the fire. When the spring is red hot plunge it, with the rod, into oil. To temper the spring, hold it with the rod over the fire, until the oil ignites and burns off.

Repeat this operation three times, then cool off in oil. This method of tempering can be applied to larger springs if necessary.

Another method of hardening spring steel is by heating to 820° C, then cooling off in oil. To temper, reheat to 380° C, and cool off in air.

FIG. 2: *To harden and temper a Die.*

Heat to a cherry-red, then plunge into water or oil. When cold, polish the die and lay on a hot bar, as shown. When the die turns dark brown, cool off in water or oil.

FIGS. 3, 4: *To harden a square or round bar.*

Dip vertically as shown, but not quickly, into the liquid.

FIG. 5: *To harden a wedge-shaped bar.*

Holding the narrow side uppermost, dip into the liquid, keeping the bar slightly sloped, as shown.

FIGS. 6, 7: To harden and temper a small drill, as shown in FIG. 6. After heating the drill to a cherry-red, plunge it into oil till cool, then polish it. Fill a metal pot with lead and heat the contents until red hot, then, to temper the drill, hold the cutting end with the tongs, and dip into the lead up to the neck, as shown in FIG. 7. Quickly withdraw the drill, and plunge it into oil to cool.

Blacksmith's Manual Illustrated

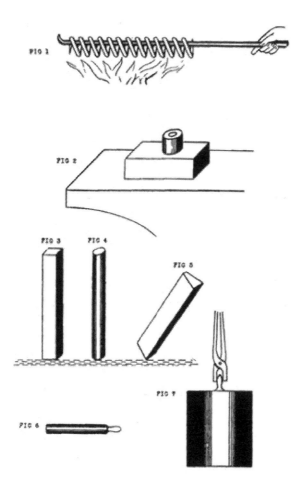

illustrates a method of hardening small drills, e.g. $\frac{1}{16}$ in.

diameter. Place the drills on a shallow tin tray, heat the tray and its contents to a dark red, and then plunge into oil to cool.

FIG. 2 illustrates the method of polishing the drills after hardening by gripping in the vice and polishing with emery paper, as shown. To temper, place on a black hot surface, keeping the drills in motion until they turn dark brown, then plunge into oil.

FIG. 3, twist drills. Heat to cherry-red, then dip vertically into the water till cool. Polish, as shown in FIG. 2. Temper by laying on a hot surface until dark brown, then plunge in oil to cool.

FIG. 4, shear blades. Heat the blade to a full red and plunge in water or oil, as shown. Polish and lay on a hot bar until the colour changes to a violet hue.

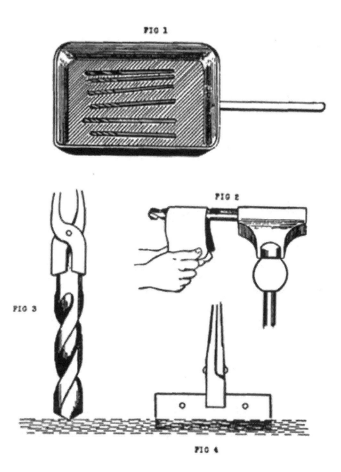

HARDENING AND TEMPERING

The process of hardening, as commonly understood by smiths, is by heating steel to a definite temperature and suddenly cooling in water or oil. This treatment causes the steel to become dead hard or glass hard, and steel in this condition is too brittle for use, therefore it has to be tempered. This can be done by heating to a certain temperature, and then cooling, or the temper can be gauged by watching the colours which appear on the surface as the heat increases. Before tempering the surface should be polished, to enable the smith to see the colours clearly. The steel is then laid on a hot surface, and as the temper increases, various colours appear on the polished surface. Immediately the desired colour is obtained, fix by plunging in water or oil. (The temperature corresponding to the different colours and shades are given in the table on temperatures and temper colours.)

Cast steel, as understood by the smith, denotes carbon steel. As regards the classification of cast steel, its carbon usually varies between ·5 and 1·5 per cent.

Steel containing ·6 to ·7 per cent, carbon is most suitable for blacksmiths' tools, such as cold chisels, hot chisels, punches, and hammers.

Points to Remember in the Treatment of Steel

Cut all tool-steel bars hot; when cut cold they are liable to crack at the end.

Always heat slowly, thoroughly, and uniformly.

When quenching tools to be hardened, keep the tool moving after immersion, thus avoiding any chance of a sharp line between the hard and soft parts of the tool.

To harden and temper a cold chisel, heat the cutting edge to a cherry-red, and immerse the chisel vertically in water. When cool, it should be slowly taken out of the water. Its internal heat will then produce the tempering colour, and this can be seen by polishing the hardened part with emery or sandstone. When the cutting edge shows the correct tempering colour (in this case purple), the chisel should be plunged at once into the water to cool. Other examples of hardening and tempering are shown on PLATES 99 and 100.

High-speed Steel

When forging high-speed steel, heat gradually to a bright red colour or between 990° C. and 1040° C, and then forge in the ordinary way. Do not continue the forging after the temperature has dropped to between 760° C. and 820° C. and the colour is below cherry-red.

ANNEALING HIGH-SPEED STEEL TOOLS

Place the tools in an iron box of sufficient size to allow at least one-half inch of packing between the tools to be annealed and the sides of the box. (Packing can consist of powdered charcoal or fine dry lime.) Cover the contents with an air-tight lid. Place the box in a furnace and heat gradually to between 760° C. and 820° C. Maintain this temperature for four hours or more, according to the quantity of steel charged, and then allow the box and its contents to remain in the furnace until cold.

HARDENING HIGH-SPEED STEEL

To harden turning and planing tools. Heat the cutting end of the tool, slowly and uniformly, to a temperature of about 760° C. or a cherry-red colour, and then bring the heat quickly to between 1250° C. and 1280° C. or a white heat, after which the tool should be cooled in a strong air-blast.

Case Hardening

Case hardening is a process of introducing carbon into the surface of low carbon steel, to harden the exterior like "cast steel" and allow the interior to retain its original properties.

The method to adopt in case hardening is as follows: Pack the articles, with a reliable casing compound, in an iron box, putting a layer of the casing compound 1½ in. deep in the bottom of the box, and laying the articles on top, leaving a 1½ in. space between each article. Place another layer of casing compound on top, and repeat as above until the box is filled to within 1½ in. from the top. Place a lining of fireclay round the inner edge of the box, and seal the box by placing a lid in the inside. This can be made airtight by placing another layer of fireclay around the edges of the lid. The box and its contents are placed into the furnace and heated to 900° C. to 950° C, and kept at this temperature until a sufficient depth of casing is obtained. If a casing $\frac{1}{10}$ in. is required, heat for about four hours. After the desired penetration has been obtained, the box should be withdrawn from the furnace and put aside to cool. When cold the articles are taken out of the box and heated to 780° C, and then quenched in cold water to obtain a refined and glass-hard casing.

Example: To convert 520° Centigrade to Fahrenheit.

$$(\cancel{520}^{104} \times \cancel{9}^{\cancel{5}}) + 32$$
$$= 936 + 32 = 968° \text{ Fahrenheit.}$$

Example: To convert 1148° Fahrenheit to Centigrade.

$$(1148 - 32) \times \frac{5}{9}$$
$$= \cancel{1116}^{124} \times 5 = 620° \text{ Centigrade.}$$

TEMPERATURE CONVERSION TABLE

Centigrade—Fahrenheit

C.	F.	C.	F.	C.	F.	C.	F.	C.	F.	C.	F.	C.	F.
38	100	63	145	88	190	165	329	400	752	720	1330	970	1778
39	102	64	147	89	192	170	338	420	788	730	1348	980	1798
40	104	65	149	90	194	175	347	440	824	740	1366	990	1814
41	106	66	151	91	196	180	356	460	860	750	1384	1000	1832
42	108	67	153	92	198	190	374	480	896	760	1402	1010	1850
43	109	68	154	93	199	200	392	500	932	770	1420	1020	1869
44	111	69	156	94	201	210	410	520	968	780	1438	1030	1886
45	113	70	158	95	203	220	428	540	1004	790	1454	1040	1904
46	115	71	160	96	205	230	446	550	1021	800	1472	1050	1922
47	117	72	162	97	207	240	464	560	1040	810	1490	1060	1940
48	118	73	163	98	208	250	482	570	1058	820	1508	1070	1958
49	120	74	165	99	210	260	500	580	1076	830	1526	1080	1978
50	122	75	167	100	212	270	518	590	1094	840	1544	1090	1996
51	124	76	169	105	221	280	536	600	1112	850	1562	1100	2014
52	126	77	171	110	230	290	554	610	1130	860	1579	1110	2030
53	127	78	172	115	239	300	572	620	1148	870	1600	1120	2052
54	129	79	174	120	248	310	590	630	1166	880	1618	1130	2068
55	131	80	176	125	257	320	608	640	1184	890	1636	1140	2086
56	133	81	178	130	266	330	626	650	1202	900	1652	1150	2102
57	135	82	180	135	275	340	644	660	1218	910	1670	1160	2122
58	136	83	181	140	284	350	662	670	1240	920	1687	1170	2138
59	138	84	183	145	293	360	680	680	1254	930	1706	1180	2158
60	140	85	185	150	302	370	698	690	1272	940	1724	1190	2172
61	142	86	187	155	311	380	716	700	1292	950	1742	1200	2192
62	144	87	189	160	320	390	734	710	1312	960	1758		

BAR STEEL WEIGHT PER LINEAL FOOT

Square.		Round.		Octagon.	
Size.	Pounds.	Size.	Pounds.	Size.	Pounds.
1/8	·05	1/8	·04	1/8	·04
1/4	·21	1/4	·17	1/4	·18
3/8	·48	3/8	·38	3/8	·40
1/2	·85	1/2	·67	1/2	·70
5/8	1·33	5/8	1·04	5/8	1·10
3/4	1·92	3/4	1·50	3/4	1·58
7/8	2·60	7/8	2·04	7/8	2·16
1	3·40	1	2·67	1	2·82
1 1/8	4·30	1 1/8	3·38	1 1/8	3·56
1 1/4	5·31	1 1/4	4·17	1 1/4	4·40
1 3/8	6·43	1 3/8	5·05	1 3/8	5·32
1 1/2	7·65	1 1/2	6·01	1 1/2	6·34
1 5/8	8·98	1 5/8	7·05	1 5/8	7·32
1 3/4	10·40	1 3/4	8·18	1 3/4	8·64
1 7/8	11·90	1 7/8	9·38	1 7/8	9·92
2	13·60	2	10·71	2	11·28
2 1/8	15·40	2 1/8	12·05	2 1/8	12·71
2 1/4	17·20	2 1/4	13·60	2 1/4	14·24
2 3/8	19·20	2 3/8	15·10	2 3/8	15·88
2 1/2	21·20	2 1/2	16·68	2 1/2	17·65
2 5/8	23·50	2 5/8	18·39	2 5/8	19·45
2 3/4	25·70	2 3/4	20·18	2 3/4	21·28
2 7/8	28·20	2 7/8	22·06	2 7/8	23·28
3	30·00	3	24·10	3	25·36
3 1/8	33·13	3 1/8	26·12	3 1/8	27·50
3 1/4	35·90	3 1/4	28·30	3 1/4	29·28
3 3/8	38·64	3 3/8	30·45	3 3/8	32·10
3 1/2	41·60	3 1/2	32·70	3 1/2	34·56
3 5/8	44·57	3 5/8	35·20	3 5/8	37·05
3 3/4	47·80	3 3/4	37·54	3 3/4	39·68
4	54·40	4	42·72	4	45·12
4 1/4	61·40	4 1/4	48·30	4 1/4	50·84
4 1/2	68·90	4 1/2	54·60	4 1/2	56·96
4 3/4	76·70	4 3/4	60·30	4 3/4	63·52
5	85·00	5	66·80	5	70·60
5 1/4	93·70	5 1/4	73·60	5 1/4	77·80
5 1/2	102·80	5 1/2	80·80	5 1/2	85·15
5 3/4	112·40	5 3/4	88·30	5 3/4	93·12
6	122·40	6	96·10	6	101·45
6 1/4	143·60	6 1/4	113·20	6 1/4	117·12
7	166·40	7	130·80	7	138·24
8	217·60	8	170·88	8	180·48
9	275·60	9	218·40	9	227·84
10	340·00	10	267·20	10	282·40
11	411·20	11	323·00	11	340·60
12	489·60	12	384·40	12	405·80

Blacksmith's Manual Illustrated

Made in the USA
Middletown, DE
09 August 2019